Engineering Materials

This series provides topical information on innovative, structural and functional materials and composites with applications in optical, electrical, mechanical, civil, aeronautical, medical, bio- and nano-engineering. The individual volumes are complete, comprehensive monographs covering the structure, properties, manufacturing process and applications of these materials. This multidisciplinary series is devoted to professionals, students and all those interested in the latest developments in the Materials Science field, that look for a carefully selected collection of high quality review articles on their respective field of expertise.

**Indexed at Compendex (2021) and Scopus (2022)**

Muhammad Ikram · Ali Haider · Sawaira Moeen ·
Junaid Haider

# Carbon-Based Nanomaterials for Environmental Applications

Springer

Muhammad Ikram
Solar Cell Applications Research Lab
Department of Physics
Government College University
Lahore, Pakistan

Sawaira Moeen
Solar Cell Applications Research Lab
Department of Physics
Government College University
Lahore, Pakistan

Ali Haider
Department of Clinical Sciences, Faculty
of Veterinary and Animal Sciences
Muhammad Nawaz Shareef University
of Agriculture,
Multan, Pakistan

Junaid Haider
Tianjin Institute of Industrial Biotechnology
Chinese Academy of Sciences
Tianjin, China

ISSN 1612-1317 ISSN 1868-1212 (electronic)
Engineering Materials
ISBN 978-3-031-59389-5 ISBN 978-3-031-59390-1 (eBook)
https://doi.org/10.1007/978-3-031-59390-1



This Springer imprint is published by the registered company Springer Nature Switzerland AG
The registered company address is: Gewerbestrasse 11, 6330 Cham, Switzerland

***Dedicated to…***
***MY PARENTS***
*For their support throughout my education as well as for teaching me to be independent and responsible*
***MY BROTHERS, SISTERS***
*For their love and support*

# Preface

The domain of nanomaterials is one of the few areas in the history of scientific research that holds tremendous promise and possibility for transformation. Within this enormous arena, carbon-based nanomaterials stand out as true innovators, offering a wide variety of applications that hold the key to tackling some of the most severe environmental concerns of our day through the formulation of solutions. As we approach a new era characterized by growing worries about pollution, climate change, and sustainability, the need to create innovative solutions is more critical than ever. The merging of nanotechnology and environmental research is crucial in this setting.

The purpose of this book is to investigate the junction of these two topics, specifically to dig into the intriguing world of carbon-based nanomaterials and the numerous uses that they have in protecting our planet. These materials have a wide range of potential applications, including the reduction of pollutants in the air and water, and antibacterial activity and many other applications.

Over the course of these pages, readers will discover an in-depth investigation of the fundamental characteristics, synthesis processes, characterization techniques, and environmental applications of nanomaterials that are based on carbon. Through cutting-edge research and practical insights, we seek to give a complete knowledge of how these materials may shape a sustainable future. We have high hopes that this book will serve as a source of knowledge and inspiration for a wide range of individuals, including scholars, students, and practitioners. We hope that by developing a more profound appreciation for the extraordinary possibilities of carbon-based nanomaterials, we will be able to stimulate new lines of investigation and invention that will move us toward a future that is greener, cleaner, and more resilient.


| | |
|---|---|
| Lahore, Pakistan | Muhammad Ikram |
| Multan, Pakistan | Ali Haider |
| Lahore, Pakistan | Sawaira Moeen |
| Tianjin, China | Junaid Haider |

# Preface

# Acknowledgments

All praises to "ALLAH," the Almighty, most Gracious, the most Merciful, and the Sustainer of the worlds, who sent us "MUHAMMAD (PBUH)" as a blessing for whole universe and the best teacher with the ultimate source of wisdom "HOLY QURAN."

First and foremost, we want to express our gratitude to ALLAH, the Almighty, for bestowing upon us the health to publish our book. Sawaira showed great pleasure to express my deepest gratitude and profound sense of obligation to my role model, Dr. Muhammad Ikram, Assistant Professor, Department of Physics, for his able guidance, enthusiasm, competent advice, constructive criticism, little castigation, constant encouragement, valuable suggestions, and painstaking supervision.

Lastly, we would like to express our gratitude to our families for the motivation, remembered us in pray and especially great patience they showed during this journey of book. "Thanks to everyone on my publishing team."

Muhammad Ikram
Ali Haider
Sawaira Moeen
Junaid Haider

# Contents

# Abbreviations

| | |
|---|---|
| 0D | Zero dimension |
| 1D | One dimension |
| 2D | Two dimension |
| 3D | Three dimension |
| AOPs | Advanced oxidation process |
| APDT | Antibacterial Photodynamic Therapy |
| BOD | Biological oxygen demand |
| CB | Conduction band |
| CBNMs | Carbon-based nanomaterials |
| CDs | Carbon dots |
| CF | Carbon fibers |
| CNTs | Carbon nanotubes |
| COD | Chemical oxygen demand |
| Cr | Chromium |
| CSs | Carbon spheres |
| CVD | Chemical vapor deposition method |
| DCFH-DA | Dichlorodihydrofluorescein diacetate |
| $e^-$ | Electron |
| *E. coli* | Escherichia coli |
| Eg | Bandgap energy |
| G +ve | Gram positive |
| G –ve | Gram negative |
| g-$C_3N_4$ | Graphitic carbon nitride |
| GNRs | Graphene nanoribbon |
| GO | Graphene oxide |
| GONWs | Graphene oxide nanowalls |
| GQDs | Graphene quantum dots |
| Gr | Graphene |
| GSH | Glutathione |
| GSSG | Glutathione disulfide |
| Gt | Graphite |

| | |
|---|---|
| GtO | Graphite oxide |
| $h^+$ | Hole |
| HCSs | Hollow carbon spheres |
| HOPG | Highly ordered pyrolytic graphite |
| LPS | Lipopolysaccharides |
| MP | Mesophase pitch |
| MWCNTs | Multi-walled carbon nanotubes |
| NF | Nanofiltration |
| NMs | Nanomaterials |
| NOM | Natural organic matter |
| $O_3$ | Ozone |
| PAN | Polyacrylonitrile |
| PC | Photocatalyst |
| PCA | Photocatalytic activity |
| PD | Photodegradation |
| PL | Photoluminescence |
| PS | Photosystem |
| PTT | Photothermal treatment |
| RGNWs | Reduced graphene oxide nanowalls |
| rGO | Reduced graphene oxide |
| ROS | Reactive oxygen species |
| SAMs | Self-assembled monolayers |
| SC | Semiconductor |
| SWCNTs | Single-walled carbon nanotubes |
| VB | Valence band |
| WHO | World Health Organization |

# List of Figures

# List of Tables

# Chapter 1
# Introduction

## 1.1 History of Carbon-Based Materials

One of the most promising scientific areas of the present day is nanotechnology, which is the study and exploitation of novel characteristics of materials at the nanoscale. Carbon-based nanomaterials (CBNMs) are among the most often utilized tools in this industry due to their special qualities, which include great mechanical strength, appealing optical features, high conductivity, and huge surface area [1]. The discussion on the evolution of CBNMs is categorized into three distinct periods: pre-1960, 1960–1985, and post-1985, as outlined in Table 1.1. In 1960, pyrolytic by chemical vapor deposition (CVD), carbon fibers (CFs) from poly(acrylonitrile), and glass-like carbon from thermosetting resins were invented, distinct from CBNMs.

Using some risky experiments, Brodie prepared GtO in 1958 to determine the molecular weight of Gt thereby, giving it the name "Graphon" [3]. Then, in 1918, Kohlschütter and Haenni published a description of the characteristics of GtO [4]. In 1924, Bernal used single-crystal diffraction observations to pursue further research on the structural characteristics of GtO [5]. As characterization developments in technology, Wallace conducted initial studies on the electrical characteristics of Gt and proposed the concept of Gr in 1947 [6]. Smally made a breakthrough discovery in the field of CBNMs development in 1985 when he found a football-shaped fullerene [7]. The world's scientists are motivated to use CBNMs by the discovery of this first carbon (C) allotropy. Gr was first used as a term for Gt single sheets by Mouras in 1987, despite the fact that theoretical calculations continued to show that stable Gt single-layer sheets did not exist [8]. In 1991, Lijima made the subsequent discovery of CNTs, a different C allotropy [9]. Over 10 years later, Geim and Novoslov conducted an incredible experiment at Manchester University wherein a single-layer Gr was synthesized in 2004 with the use of a Scotch tape approach. Thus far, the several carbon forms identified such as—fullerene, CNTs, Gr, and 3D Gt. Research done by Xu et al. in 2004 on carbon dots (CDs) served as the foundation by Ponomarenko and Geim in 2008 for the synthesis of graphene quantum dots (GQDs) [10, 11].

 
M. Ikram et al., *Carbon-Based Nanomaterials for Environmental Applications*, Engineering Materials, https://doi.org/10.1007/978-3-031-59390-1_1

**Table 1.1** Short history of development of CBNMs [2]

| Period | Years | Development of CBNMs | Remarks |
|---|---|---|---|
| I | –1960 | Activated carbon, natural diamond, graphene (Gr), graphite oxide (GtO), artificial Gt blocks, carbon blacks | Items are available for purchase in either metric tons or kilograms |
| II | 1960–1985 | Several carbon fibers, graphite (Gt), synthetic diamond, carbon composites, diamond-like carbon, glass-like carbons | Sales mostly in grams |
| III | 1985– | Graphene quantum dots (GQDs), fullerenes, carbon nanotubes (CNTs), carbon dots | Nano-sized sale in milligram |

Because the Gr lattice within the GQDs is less than 100 nm and has fewer than 10 layers in thickness, they vary from CDs [12]. Due to their quantum confinement effect, GQDs have several unique qualities, including their distinctive fluorescence properties, which were found by Pan et al. in 2010 [13]. In 2012, Zhao and colleagues conducted an investigation into the alteration of GQD characteristics by doping. The study involved the use of nitrogen as a dopant in GQDs (Fig. 1.1) [1].

**Fig. 1.1** Short history of growth of CBNMs. Reproduced with permission from Ref. [1] Copyright 2018, Elsevier

## 1.2 Classification of Carbon-Based Materials

In the past few years, the world has witnessed a considerable increase in various kinds of high-quality novel CBNMs and synthesis processes due to the growing need for carbon NMs in industry [14]. There are numerous approaches to categorize CBNMs because they range greatly in terms of size, form, content, etc. [15].

### *1.2.1 Graphene*

Gr is an allotrope of the C, atomic-scale, two-dimensional, hexagonal structure composed of one atom at each vertex by $sp^2$ hybridization (Fig. 1.2a). Each lattice contains three strong σ bond connections that form a sturdy hexagonal structure. The π-bond, which is positioned vertically to the lattice plane, is primarily responsible for Gr's electrical conductivity. The closely spaced C atoms in Gr and the hybridization of the $sp^2$ orbital, which is a combination of S, Px, and Py that form the σ-bond, are the factors that contribute to the stability of Gr. The last $e^-$ of Pz generates the π-bonds and hybridization of these bonds forms the π-band and π*-bands. The majority of Gr's prominent electrical characteristics are a result of these bands, including the half-filled band that allows $e^-$ to move freely. Gr contains Gt, fullerene, and CNTs as well as aromatic molecules of infinite sizes such extremely planar polycyclic aromatic hydrocarbons. In simpler terms, Gr is made up of a single layer of C atoms arranged tightly in 2D honeycomb lattice structure. Each C atom in single-layer Gr provides an unpaired $e^-$ to a benzene ring formed by $sp^2$ hybridization. Gr has an average thickness of only 0.35 nm or 1/200,000th of a human hair's diameter. However, Gr structure is extremely stable. To avoid atom reconfiguration, the bond among C atoms is sufficiently robust to resist external stress across a twisting lattice plane [16, 17]. A stacking of Gr layers results in Gt, which has d-spacing of 0.335 nm (Fig. 1.2b) [18]. Stacking of Gt results in the formation of a 3D structure, whereas Gr is a 2D materials consisting of a single layer of atoms [19].

In addition, Gr is a naturally occurring semi-metal with a bandgap energy (Eg) separating its six π-orbitals, three of which are occupied bonding orbitals and three of which are vacant antibonding orbitals inside its electronic structure resembling benzene ring. Since there is some overlapping between the valence band (VB) and conduction bands (CB) as a result of these benzene rings fusing, $e^-$ from the upper position of VB move toward to the lowest position of CB without producing heat (Fig. 1.2c) [20, 21]. The particular surface area is determined by the accumulation of Gr layers. The effective surface area grows and aggregation is reduced when additional chemicals are added to pure Gr [22, 23].

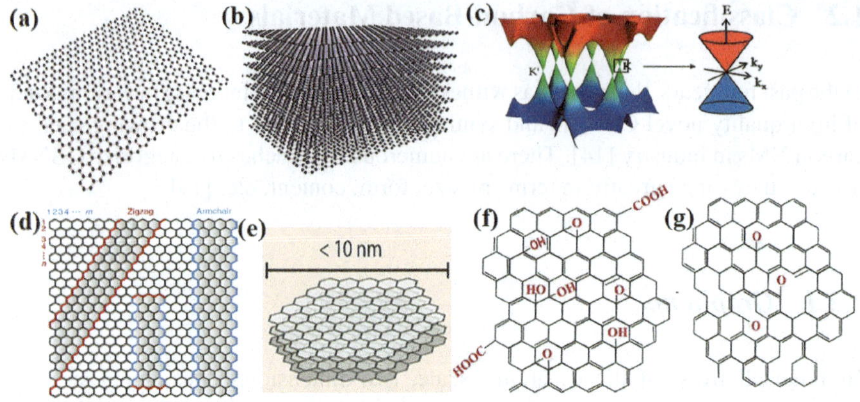


**Fig. 1.2** **a** Structure of Gr, **b** Gt, **c** electronic structure of graphene. Reproduced with permission from Ref. [31] Copyright 2015, Springer Nature **d** graphene nanoribbon Reproduced with permission from Ref. [32] Copyright 2018, Elsevier, **e** GQDs, Reproduced with permission from Ref. [33] Copyright 2019, Wiley, **f** GO and **g** rGO, Reproduced with permission from Ref. [34] Copyright 2018, Elsevier

#### 1.2.1.1 Graphene Nanoribbon (GNRs)

GNRs, a material with large aspect ratio, can be formed by the cutting of the Gr sheet. The finite, planar and quasi-1D Gr structures are known as GNRs. The structure of their longitudinal edges, which might have a zigzag, armchair, or intermediate character, is typically used to identify them (Fig. 1.2d). This edge structure has significant effects on the GNRs' electrical characteristics. GNRs are typically semiconducting, but the edge geometry has a crucial role in determining the 1D band structure and, in particular, the bandgap's magnitude. The bandgaps that differentiate GNRs from extended, 2D Gr, along with their significant charge carrier mobility, make them appealing as substances for a range of applications, including field-effect transistors (FETs) and radio-frequency transmitters [24].

#### 1.2.1.2 Graphene Quantum Dots

Zero-dimensional (0D) pieces of Gr sheets with a lateral size smaller than 100 nm are known as GQDs. Researchers are very interested in GQDs because of their intriguing physicochemical features; these particles have a structure that is comparable to Gr (Fig. 1.2e). Nanometer-sized Gr cluster $sp^2$ domains are often broken up to produce GQDs [25, 26]. GQDs have enriched edge defects, intense photoluminescence (PL), nonzero Eg, and excellent water solubility due to quantum confinement phenomena at the nanoscale. Increased $e^-$ transfer rates at the edges, little cytotoxicity, superior chemical stability, and ease of production are further benefits of GQDs. Due to the unique characteristics of GQDs; high-performance photonic; energy-related devices;

and electrical devices such as flexible memory systems, rechargeable batteries, solar cells, photodetectors, LEDs, and supercapacitors may be developed [27].

#### 1.2.1.3 Graphene Oxide (GO)

An alternate approach to achieving many of the benefits of pure Gr while simultaneously endowing the surface with functionalized oxygen groups is to create Gr-like compounds from Gt or other C sources via a top-down process (Fig. 1.2f). GtO, made up of many stacked layers of GO, is produced when Gt is oxidized in protonated fluids. They have similar structures to Gr but also comprise hydroxyl, alkoxy, carbonyl, carboxylic acid, and other oxygen-based functional groups. In addition to being simpler to synthesize, these oxygenated groups offer other benefits over Gr, such as increased solubility and surface functionalization potential, which have opened up a variety of applications for composite materials [28, 29]. Reductive exfoliation procedures can be used to restore Gr-like characteristics to GO that convert reduced graphene oxide (rGO) from GO (Fig. 1.2g). Now, rGO is starting to emerge as a very strong Gr and GO compromise. This is due to the fact that rGO not only possesses Gr-like characteristics, such as reasonably strong conductivity, but it is also simple to synthesize in appropriate amounts from inexpensive GO utilizing a range of (electro)chemical, photo-assisted, microwave, and thermal techniques [30].

### *1.2.2 Carbon Nanotubes*

Rolling up sheets of Gr creates the tubular shape of CNTs, which are $sp^2$ nanocarbon compounds. CNTs have been a special research focus in the field of carbon NMs ever since they first emerged and described. The advancements in their industrial macro-scale synthesis and associated application domains have taken center stage in current research, which has expanded beyond the investigation of their physical and chemical characteristics [35]. CNTs may be produced using three primary methods: CVD, laser ablation, and arc discharge. The chosen precursor materials, preparation techniques, and synthesis circumstances all have a significant impact on the CNTs characteristics. Although CVD is thought to be the most efficient and productive process, arc discharge and laser ablation will produce more contaminants [1]. The atomic structure of CNTs can be explained by the means of tube chirality and chiral angle. This can be seen in Fig. 1.3a when the tube is rolled until the tip of the chiral vector hits its tail and the Gt sheet is sliced along the dotted lines. The following expression describes the chiral vector $C_h$, sometimes referred to as the roll-up vector Eq. (1.1):

$$C_h = na_1 + ma_2 \tag{1.1}$$

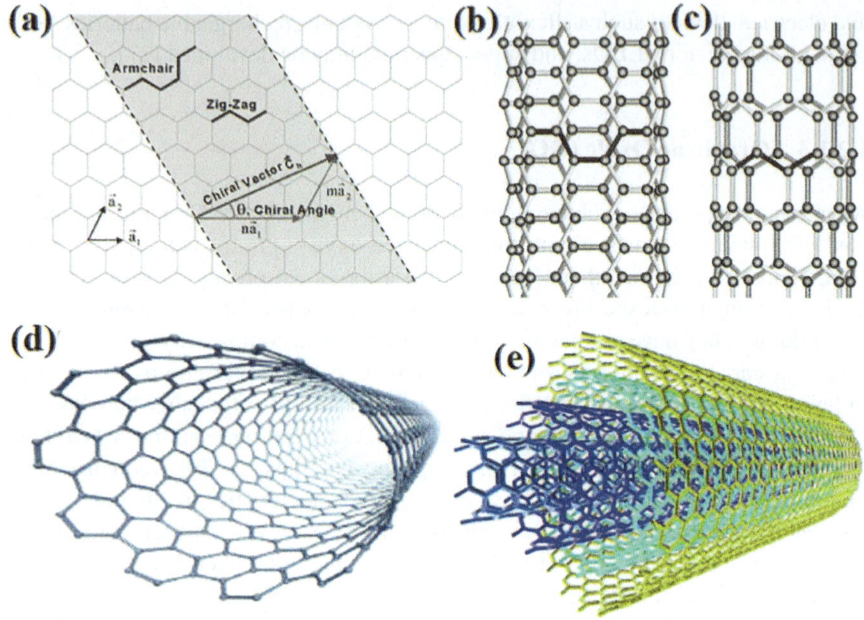


**Fig. 1.3** **a** Formation of CNTs by hexagonal Gt sheet, **b** atomic structure of an armchair, **c** atomic structure of zig-zag CNTs. Reproduced with permission from [36], Copyright 2021, Elsevier, **d** SWCNTs and **e** MWCNTs. Reproduced with permission from [50], Copyright 2016, polymer-plastics technology and engineering

where n and m are step numbers in direction of zig-zag C bond, and $a_1$ and $a_2$ are unit vectors of hexagonal lattice. The degree of "twist" in the tube is measured by the chiral angle. There are two limiting instances: 0 and 30° for the chiral angle. These limiting examples are named armchair (30°) and zig-zag (0°) according to the shape of the C bonds around the CNTs [36]. Figure 1.3b–c illustrates the distinction between the armchair and zig-zag nanotube structures. The characteristics of the material are significantly affected by the chirality of the CNTs [37]. Specifically, it is well known that the electrical characteristics of CNTs are significantly influenced by the tube chirality. Although Gt is considered as a semi-metal, research has demonstrated that, based on the tube's chirality, CNTs can have either metallic or semiconducting behavior. There have also been studies published on how chirality affects the mechanical characteristics. Yakobson et al. investigated the instability of CNTs below linear response by their analytical study [38, 39]. Their models demonstrate the extraordinary resilience of CNTs, which can withstand high strain without exhibiting any symptoms of brittleness or flexibility. Despite the fact that chirality has a negligible effect on elastic rigidity, they came to the conclusion that a major factor in the plastic distortion of nanotubes under tension is the Stone–Wales modification, a reversed diatomic exchange that results in the formation of two pentagons and two heptagons in pairs. CNTs can be classified as single-walled carbon nanotubes

(SWCNTs) or multi-walled carbon nanotubes (MWCNTs) based on whether they are made of one layer or multiple C atomic structures [40].

#### 1.2.2.1 SWCNTs

Ijima and Ichihashi added the catalyst to the Gt electrode using the arc discharge technique in 1993 to create SWCNTs [41]. SWCNTs are made up of monolayer Gr, which has 1.4 nm of average diameter, 1–2.4 nm of outer diameter, and 2–5 μm of length range (Fig. 1.3d) [42]. Its structure will become unstable as the diameter rises over 6 nm. The tiny size, strong modulus, and significant electrical conductivity of SWCNTs are quite concerning [43]. By stacking π–π connections, CNTs facilitate the efficient loading of numerous molecules along the wall. Simultaneously, the associated characteristics may be enhanced by surface chemical alteration of their interior and exterior surfaces [44]. As a result, SWCNTs offer tremendous opportunity as polymer materials in a variety of industries, including biomedicine, electronics, and manufacturing. The side edges of CNTs are very hydrophobic and hardly soluble in most solvents [45]. To solve the defects of CNTs, Liu et al. working for year and suggested the covalently combining hydrophilic compound (polyethylene glycol) with oxidized SWCNTs [46]. These SWCNTs may be non-covalently bonded to a wide range of aromatic compounds for additional practical study, and they can react to pH activation. Shape memory substances may instantly return to their previous shape in response to outside stimuli like heat and light [47]. Their cheap cost, simple processing, and responsive qualities have garnered a lot of interest. Lee and Yu combined SWCNTs with polyurethanes (less molecular weight) to create electroactive shape memory materials at less temperature. It is discovered that SWCNTs substantially improve the composite's efficiency [48].

#### 1.2.2.2 MWCNTs

The multilayered sheets of Gr that make up MWCNTs are coiled along the central axis. Typically, there are 2–50 layers, with a spacing of around 0.34 nm between each layer and an outside diameter ranging from 2.5 to 100 nm (Fig. 1.3e) [42]. Moreover, MWCNTs are among the most widely used CBNMs because of their superior mechanical and photothermal characteristics. Furthermore, MWCNTs have stronger rigidity, strength, and have a lower manufacturing expense than SWCNTs. The industry can make them in big quantities, which is very valuable for growth. Hyperbranched polyamine functionalized MWCNTs gel was created by You et al. via a multistep Michael addition procedure. When exposed to varying intensities of stimuli, this gel may react to ultrasonic waves and quickly transition between gel and sol. The composites of poly(vinylidene fluoride) and MWCNTs were studied by Dang et al. in relation to stimulus of pressure. After bending, the composites' electrical conductivity and dielectric constant consistently dropped. The material's susceptibility to tensile stimuli will also be impacted by the MWCNTs concentration [49].

### *1.2.3 Carbon Fibers*

Carbon fibers (CF) are fibers with a tremendous amount of strength that are formed when C atoms are bonded together in 1D. Petroleum pitch and rayon, particularly the higher modulus types, account for a tiny portion of CF, with polyacrylonitrile (PAN) accounting for around 90% of the total. A significant portion of the expense, almost 50%, of the cost of creating CF goes on generating the precursor PAN fiber. Therefore, in order to reduce the cost of generating CF and gain traction in large volume application sectors such as automotive where cost is a key concern, emphasis has recently been directed toward other, more affordable, and environmental precursors such as lignin, polyethylene, or cellulose [51–54]. The atomic structure of CF is composed of uniform hexagon-shaped layers of C atoms, akin to the structure of Gt. CF layer planes can have a hybrid, turbostratic, or graphitic structure, based on the precursors and synthesis techniques used. The layer planes in graphitic areas are regularly arranged parallel to one another. Van der Waals forces, which are very weak, interact among the sheets, but atoms in a plane are covalently bound by means of $sp^2$ bonding. The distance ($d_{002}$) among the two Gr layers in a single graphitic crystal is approximately 0.335 nm [55]. $C_{44}$ for shearing is as small as 4.5 GPa, whereas $C_{33}$ and $C_{11}$ are 36.5 and 1,060 GPa, respectively. Nonetheless, a stack of turbostratic layers makes up the fundamental structural component of many CF. The adjacent Gr sheets of a turbostratic structure are folded, twisted, or split randomly, or they are piled erratically. According to certain reports, $sp^3$ bonding and uneven stacking can cause d-spacing to grow to 0.344 nm [56, 57]. In general, CF from mesophase pitch (MP) and vapor produced CF are the only ones that exhibit a graphitic crystalline structure; CF from other precursors, such as PAN, can have a turbostratic structure. Crystallites arise in the graphitization of fixed PAN fibers by a combination of integrating adjacent disordered C or consolidating with neighboring crystallites. Furthermore, the layer planes inside the crystallites reorganize by rotating and moving [58].

#### 1.2.3.1 PAN-Based CF

PAN, including approximately 68 wt% C, the primary precursor used in the synthesis of CF. The existence of polar nitrile groups provides strong intermolecular interactions, raising PAN's melting point. PAN is mostly processed using solution spinning methods since it has a tendency to degrade at temperatures lower than its melting point. Incorporating acidic co-monomers reduces the glass-transition and stabilization temperatures, facilitating the release of heat and enhancing the advancement of stabilization processes. The process of manufacturing CF requires many stages, such as stabilization, graphitization, polymer synthesis, carbonization, and fiber spinning

[59–61]. The stabilization reaction entails cyclization, wherein PAN's nitrile groups change into a structure resembling a ladder that is resistant to high temperatures and increases the amount of C produced during the carbonization process. Moreover, oxidation processes take place during high-temperature treatment, significantly improving the material's thermal resilience. Furthermore, dehydration reactions take place, resulting in the formation of C=C bonds and the release of $H_2O$ molecules. The carbonization procedure entails the elimination of oxygen and other elemental components, while simultaneously enhancing the crystallinity of the CF by boosting the bonding between C atoms. The graphitization process reorganizes the C atoms into a hexagonal arrangement, resulting in the formation of highly crystalline fibers with a high modulus. By meticulously optimizing the production conditions, the tensile characteristics of the resulting CF may be improved [62]. The gel-spinning approach is now the most often used method for producing precursor fibers [63–65].

#### 1.2.3.2 Pitch-Based CF

Natural pitch, consisting of 80% C, is derived from the process of destructive distilling coal and petroleum, whereas synthetic pitch is obtained by the pyrolysis of synthetic polymers. Coal pitch often exhibits a higher level of aromaticity compared to petroleum pitch. According to Smith et al., around two-thirds of the chemicals found in coal tar pitch were aromatic, while the other compounds were heterocyclic [66]. Nevertheless, coal pitch frequently contains a substantial amount of C particles (solid matter), leading to filament rupture during extrusion and heat processing. Hence, petroleum pitch is the preferable choice for CF synthesis. The commercial pitch comprises aromatic compounds with molecular weights ranging from 400 to 600 [67]. The recent surge in interest in synthetic pitch can be attributed to its superior purity and its ability to stabilize more rapidly at a given temperature. Pitch is a broad word that refers to a dense, elastic, black substance often derived from coal- and petroleum-based materials [68]. In ancient times, natural pitch was utilized in shipbuilding, particularly in ancient Egypt, to make wooden boats watertight. These boats were constructed by joining wooden planks together with wooden pegs [69]. Pitch obtained from petroleum products is a cost-effective substitute for pitch made from coal tar, as it possesses similar physical qualities [70]. The elastic modulus of pitch is higher than that of PAN due to pitch's greater graphitizability compared to PAN. Pitch, generated from petroleum, is a viscous and dark substance that is obtained by the process of fractionation and distillation of crude oil [71]. Coal pitches exhibit a higher level of aromaticity compared to petroleum pitches due to the solid C particles' capacity to accelerate the formation of coke during extrusion and thermal processing. Despite their reduced aromaticity, petroleum pitches remain appealing as precursors for CF.

#### 1.2.3.3 Lignin-Based CF

Biopolymers are being studied as precursors for CF, along with PAN and pitch. During the initial stages of CF development, rayon served as the prevailing precursor. However, its restricted availability was hindered by poor C yields and high manufacturing costs. Biopolymers should typically have a high C content and allow for massive C production at low costs. Lignin is a component of wood and is acquired as a substantial residue from chemical pulping processes in the paper industry. Lignin is the sole biopolymer that possesses significant amounts of aromatic compounds, namely, phenolic groups, in its structure. The Kraft process is the predominant method for extracting lignin, whereby sulfides and alkali salts are employed to break down the native lignin into soluble oligomers, facilitating its separation from rayon. In addition to Kraft lignin and other lignin methods, the commercially accessible supply of organosolv lignin is significantly restricted compared to Kraft lignin.

Lignin was first studied as a CF precursor in the late 1960s [72]. Sudo et al. made modifications to lignin in the 1990s by introducing hydroxyl groups, resulting in the production of CF with moderate characteristics [73, 74]. Sano and his colleagues used melt-spinning to study pure organosolv lignin, which was sulfur-free. The CF they developed exhibited much reduced tensile strengths, measuring at 0.3 GPa, in contrast to a typical commercially available PAN-based CF that begins at 3.5 GPa [75]. The organosolv and Kraft lignin as well as mixture of lignin were examined by Kadla et al. The melt-spun fibers were subjected to a stabilization procedure at a temperature of 250 °C, and then carbonized at 1000 °C. As a consequence, the tensile strengths reached around 0.5 GPa and the moduli reached a maximum of 60 GPa [76]. PET combined with Kubo and Kadla's blending process produced fibers with 0.7 GPa tensile strength and 94 GPa modulus [77]. The Oak Ridge National Laboratory (ORNL) studied hardwood lignin and softwood–hardwood mixtures [78]. A recent publication from ORNL discusses the development of CF produced from lignin, which have tensile strengths of around 0.5 and a modulus of 29 GPa [79]. A report from ORNL demonstrates that CF, when carbonized at a temperature of 1200 °C, have tensile strengths of around 1.0 GPa [80].

So far, all the published data on CF made from lignin have been acquired through studies that were not continuous and did not allow for stretching or continuous fiber processing. The available methods for stabilizing published materials are primarily restricted to a gradual heating process in air, at a rate of 0.5 °C $min^{-1}$. Therefore, it is important to acknowledge that the current data is not as optimized as the well-researched PAN- and pitch-based precursors [81].

### *1.2.4 Carbon Spheres*

Various nomenclatures have been assigned to CBNMs with a spherical form. These encompass many forms of C such as carbon blacks, mesoporous microbeads, carbon balls, carbon microbeads, onions, carbon spheres (CSs), and so on. Several C structures have been seen to consist of concentric layers that are often fragmented and extend outward from the center. CNTs are drawn together by van der Waals forces, causing the formation of agglomerated clusters of CNTs. When CSs have diameters lesser than 1000 nm, they likely to undergo accretion, which means they join together to create structures resembling necklaces or beads. Therefore, whereas CSs are typically regarded as distinct entities, they frequently assemble into interconnected, continuous chains. These chains are generated at an early stage of the synthesis process and are not always necessary owing to post-synthesis treatment. CSs can also be prepared by arranging C layers that are oriented perpendicular to the rim of the sphere. The study of CSs may be classified into different categories. Firstly, spheres can be categorized as either solid, core–shell, or hollow structures [82]. Additionally, they may be categorized based on their "nanometric texture," specifically in relation to whether the spheres consist of radial, concentric, or random layers. Furthermore, there is a proposition that suggests spheres may be classified based on their size, namely, their diameter [83].

This approach classifies particles into three categories: well-graphitized spheres with sizes ranging from 2 to 20 nm, less graphitized spheres with sizes ranging from 50 to 1000 nm, and C beads larger than 1000 nm. Additionally, spheres can be categorized based on the methodologies employed in their synthesis. The many methods employed in the production of spherical CBNMs, including as laser ablation, CVD, autoclave techniques, and arc discharge, provide materials with distinct characteristics. CSs range from a few nanometers in onions to several microns during polymerization.

Typically, during a synthesis method, a diverse set of diameters is synthesized. The smaller CSs, normally produced by CVD techniques, are commonly observed in the shape of a necklace, where the spheres are connected (aggregated) in a 2D strand that can span over many spheres. At the first stage of the synthesis process, the existence of sphere connections can be detected, wherein the spheres are surrounded by Gt layers forming a structure resembling a necklace. Generally, bigger spheres are formed as distinct structures made of C that can interact with other spheres by van der Waals bonding. Figure 1.4a, b illustrates the impact of heating individual CSs (a) in order to create fused structures through the merging of spheres (b), which occurs at elevated temperatures. An arc discharge process using C balls as electrodes produces MWCNTs via high-temperature accretion, according to research [84]. A theoretical framework incorporating imperfections and heptagon/pentagon arrangements was employed to elucidate the process of conversion. Theory and experiment have been published on spherical objects (oil droplets covered with silica/CTAB) that accrete to construct structures.

These investigations indicate that C necklaces might be manufactured with precise shapes and dimensions [85]. As previously stated, spheres can be organized with their C chains/flakes/layers positioned either at a right angle or in alignment with C core (Fig. 1.4c). The configuration of the layers has a substantial influence on the chemical and physical properties of the CSs. Figure 1.4d depicts the alignment of C flakes adjacent to with the C core. The flakes typically have diameters ranging from 2 to

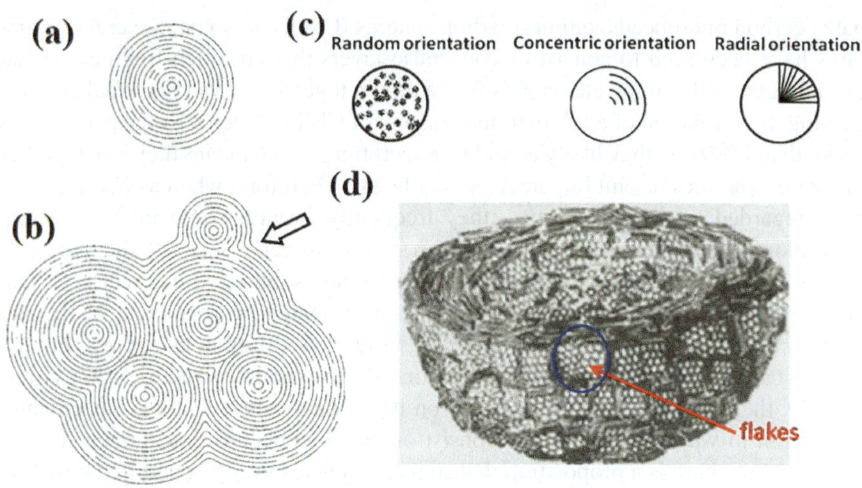


**Fig. 1.4** **a** Structure of C onion; **b** aggregation of C onion resembling structure (arrow designates 'neck' among carbon onions); **c** radial, concentric, and random C layer alignments in CSs; and **d** illustration of a cross section of a solitary C black particle, displaying concentric layers. Reproduced with permission from [105], Copyright 2010, Elsevier

10 nm [86]. The size of the flakes can be determined by TEM or XRD investigations. The pyrolysis of the CSs results in the development of a more graphitic structure, characterized by an increase in the size of the flakes.

#### 1.2.4.1 Hollow Carbon Spheres

The hollow carbon spheres (HCSs) have low density, extensive pore structure, chemical stability, and high specific surface area. Inside the hollow, drug loading is possible due to its 8 nm diameter. HCSs may be fabricated using many general methods, with the templating technique being the most often utilized. The template technique involves the encapsulation of a spherical material, such as a metal complex, micelle, or silica, with a C-containing substance. After self-assembling, this CBNMs forms a core–shell structure with a filled interior by forming C–C bonds on the sphere. The sphere's inside can include elements with exploitable qualities. For instance, if the core is made of a magnetic material, the CSs can be changed using magnetic fields [87, 88]. Alternatively, the core can be extracted, often using heat means, in order to produce a hollow spherical. The template is thus forfeited during the synthesis. Polyelectrolyte capsules and similar substances also enable the creation of empty spheres. However, in this approach, the development occurs inside the cavity rather than in the outside layer, resulting in parachute-like structures [89]. A recent paper discussed the production of hollow inorganic spheres by temperature control during synthesis. The technique facilitated the formation of vanadium oxide spheres with two or possibly several layers [90]. The suggested approach aims to achieve the formation of structures through precise regulation of chemical reactions. The resulting materials are expected to exhibit exceptional thermal stability. This approach has consequences for creating stacked multi-walled C structures.

HCSs have favorable characteristics such as significant chemical inertness, excellent electrical conductivity, and adjustable surface properties, making them very promising for applications as electrode materials, biomaterials, catalyst supports, and photonic Eg materials [91]. Several techniques have been shown for the preparation of HCSs. The technique of nanocasting solid mesoporous/core–shell silica spheres resulted in the creation of C capsules that had a hollow center and a shell that possessed mesoporous features [92]. Additionally, HCSs were manufactured by using macroporous silica as a reference material [93]. The development of porous hollow C structures was made feasible by the utilization of a template consisting of a spherical core–shell structure ($SiO_2@ZrO_2$) [94]. The sol–gel approach was used to synthesize tin-encapsulated spherical HCSs. A recent study demonstrated the synthesis of graphitic HCSs utilizing cellulose chars, resorcinol–formaldehyde gel, and polypyrrole. Furthermore, the process involved the utilization of oxidized fullerenes to construct HCSs. Additionally, several other techniques were documented, including the self-assembly method medial-reduction, carbon dioxide method, and co-pyrolysis methods. Regrettably, none of the previously suggested techniques provide exact control over the shell thickness, core diameter, composition, and overall shape of the resulting HCSs [91].

Heteroatoms, namely nitrogen, have been utilized to introduce into the carbon Gr layer in order to accurately modify the physicochemical properties of the CBNMs. These properties encompass attributes like adsorption ability, crystallinity, electrical conductivity, and chemical reactivity [95–98]. The template approach has been employed to synthesize microporous C, N-doped CNTs, and organized mesoporous carbon. Wang et al. synthesized a composite material by integrating $TiO_2$ with HCS. The resultant compound exhibited the capacity to sequester carbon dioxide upon exposure to UV irradiation. The presence of a hollow C NSs facilitates numerous scattering of light within the composite material, resulting in a significant enhancement of light absorption efficiency. The composite material has a significantly elevated specific surface area, resulting in a remarkable improvement in the carbon dioxide adsorption efficiency [99].

#### 1.2.4.2 Porous Carbon Spheres

Non-porous C structures have inferior performance in chemical loading, but the permeability of porous C structures enables them to exhibit outstanding performance. CBNMs possess a porous structure, which enables them to be utilized in many fields such as energy, environmental protection, catalysts, water treatment, $CO_2$ collection, and energy conversion [100, 101]. Kapri et al. produced porous CSs with a diameter of 150 nm using cymbopogon flexuosus as the source material [102]. Subsequently, they modified the CSs by introducing compounds, such as doxorubicin (DOX), to confer pH-responsive characteristics. Research has been conducted using porous C NPs to enhance the stimulus response performance by modifying them. Zhang et al. coated CSs with mesoporous CuS nanoparticles [103]. The CuS NPs possess NIR absorption and photothermal conversion capabilities, enabling them to respond to light. The resulting heat can promote the liberation of loaded chemicals, such as DOX.

The varied stimulus response of several CBNMs is investigated through comparative tests. Zhou and colleagues synthesized light-responsive mesoporous CSs using a silica-assisted synthesis technique. Mesoporous CSs have high absorption efficacy across the UV-Vis-NIR optical range, spanning from 300 to 1,400 nm. Research has shown that in the near-infrared regions NIR-I and NIR-II, mesoporous CSs have an absorption coefficient that is 1.5 times greater than that of SWCNTs and Gr. Additionally, mesoporous CSs have a better photothermal conversion efficiency [104].

### *1.2.5 Graphitic Carbon Nitride*

Carbon nitride ($C_3N_4$) is a polymer semiconductor (SC) material that consists of a binary compound of C and nitrogen [102]. The nitrogen lone pair interaction with states gives $C_3N_4$ its distinctive structure, making it an extremely promising material for various of applications [106, 107]. In 1989, Liu and Cohen synthesized a compound called β-carbon nitride (β-$C_3N_4$) by substituting C for silicon in β-$Si_3N_4$ [108]. The study discovered that β-$C_3N_4$ is a covalent crystal. Through ab initio calculation, it was projected that β-$C_3N_4$ is a rigid material that is comparable to, or maybe even surpasses, the hardness of diamond. Since then, $C_3N_4$ has garnered global scientific attention because of its exceptional qualities, including a low friction coefficient, dependability, chemical inertness, and hardness. Corkill and Cohen determined the Eg of β-$C_3N_4$ to be $6.4 \pm 0.5$ eV [109]. Subsequently, Liu and Wentzcovitch identified two further structures, namely, cubic carbon nitride and layered graphitic-like rhombohedral $C_3N_4$ [110]. Hemley and Teterput identified five different structures of $C_3N_4$, which are α-$C_3N_4$, β-$C_3N_4$, pseudo-cubic $C_3N_4$, cubic $C_3N_4$, and g-$C_3N_4$. These structures had Eg of 5.49, 4.85, 4.13, 4.30, and 2.88 eV, respectively, as shown in Fig. 1.5 [111].

Out of these five structures, g-$C_3N_4$ was shown to be the most stable form of $C_3N_4$, with the lowest energy. The Eg of g-$C_3N_4$ is approximately 2.7–2.8 eV. The Eg provides a flexible method for controlling the HOMO (highest occupied molecular orbital) and LUMO (lowest unoccupied molecular orbital) [112]. The chemical g-$C_3N_4$ is present in three distinct forms: a hexagonal structure known as s-triazine-bases, a structure derived from tri-s-triazine (heptazine), and an orthorhombic structure derived from s-triazine (Fig. 1.6) [106]. The g-$C_3N_4$ structure derived from tri-s-triazine was shown to exhibit the highest stability compared to other forms of g-$C_3N_4$. The thermal stability of g-$C_3N_4$ has been verified by TGA, showing resistance

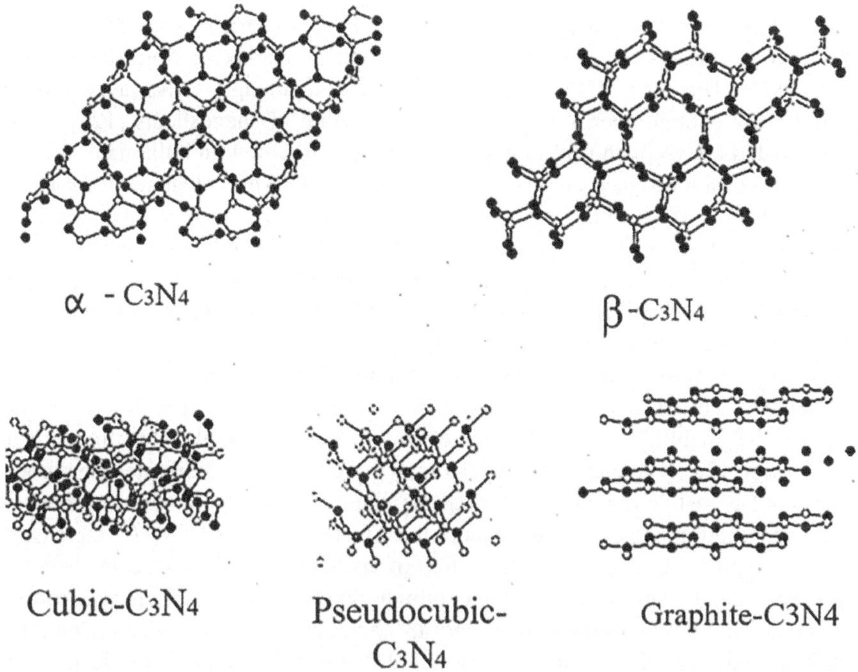


**Fig. 1.5** Structure of α-$C_3N_4$, β-$C_3N_4$, cubic-$C_3N_4$, pseudocubic-$C_3N_4$, and g-$C_3N_4$. Reproduced with permission from [126], Copyrights 2023, Elsevier

to temperatures up to 600 °C. It exhibits chemical durability and is resistant to dissolution in acid, alkali, or organic solvents, which verifies its outstanding stability under normal environmental circumstances [113]. The chemical is named g-$C_3N_4$ due to its structural resemblances to graphitic nanostructures (NSs). The material exhibits exceptional stability, a high refractive index, excellent thermal conductivity, remarkable frictional capabilities, high toughness, distinctive electrical features, and notable chemical properties [114, 115]. Ever since its discovery in 2009, g-$C_3N_4$ has been widely recognized as a very promising photocatalyst (PC) compared to many other SCs [106]. The bulk morphology of g-$C_3N_4$ is characterized by a three-dimensional arrangement and shares certain resemblances with Gt. Typically, large quantities of g-$C_3N_4$ are created by the process of pyrolysis using nitrogen-rich compounds such as thiourea, melamine, urea, and dicyandiamide [116]. Despite possessing a 2.7 eV Eg, bulk g-$C_3N_4$ has a lower specific surface area due to layer stacking during polycondensation. Due to limited or nonexistent electronic transition, it has a poor quantum yield and a fast photogenerated $h^+$ and $e^-$ recombination rate [117]. In order to overcome these limitations, several efforts have been undertaken to enhance its photocatalytic activity (PCA). The technique involves breaking down vast amounts of g-$C_3N_4$ into CNTs, nanosheets, and QDs by exfoliation. Furthermore, it entails the introduction of supplementary compounds into g-$C_3N_4$ by doping, or the fusion of g-$C_3N_4$ with other materials to generate nanocomposites [115, 118–120]. Bulk

$g\text{-}C_3N_4$ has a layered, planar structure like Gt [121]. The layers are attracted by small van der Waals forces despite covalent connections between nitrogen and C atoms [122]. The van der Waals forces of interaction among $g\text{-}C_3N_4$ layers can be disturbed by enough energy in a solvent to generate 2D nanosheets [123]. Yang et al. created $g\text{-}C_3N_4$ nanosheets via liquid-phase exfoliation on bulk material. The nanosheets developed have an Eg value of 2.65 eV [124]. Niu et al. generated $g\text{-}C_3N_4$ nanosheets by exposing bulk $g\text{-}C_3N_4$ to thermal oxidation etching in an ambient air environment. The $g\text{-}C_3N_4$ nanosheets that were produced exhibited a significant Eg of roughly 2.9 eV, a specific surface area more than 306 $m^2\ g^{-1}$, and enhanced electron transport in comparison to bulk $g\text{-}C_3N_4$. The considerable improvement can be attributed to the substantial quantum confinement [125].

The practical utility of bare $g\text{-}C_3N_4$ is limited due to certain impediments and deficiencies. The SC's PCA is hindered by its limited surface area, low electrical conductivity, inability to absorb light beyond 460 nm, and high recombination of $e^-$–$h^+$ pairs [127]. The interconnected structure of conjugation facilitates the recombination of $e^-$, while the modification of the material can prevent the movement of $h^+$. Two methods are suggested to enhance the effectiveness of $g\text{-}C_3N_4$: (i) decreasing the Eg of the PC SC to improve absorption of visible light and (ii) effectively separating excitons to minimize $e^-$–$h^+$ recombination during the catalytic reaction by introducing defects and local lattice. Various approaches have been documented for obtaining changed $g\text{-}C_3N_4$, including modified synthesis techniques, NSs design, and electronic structure manipulation [119, 128]. Infecting, copolymerizing, and doping

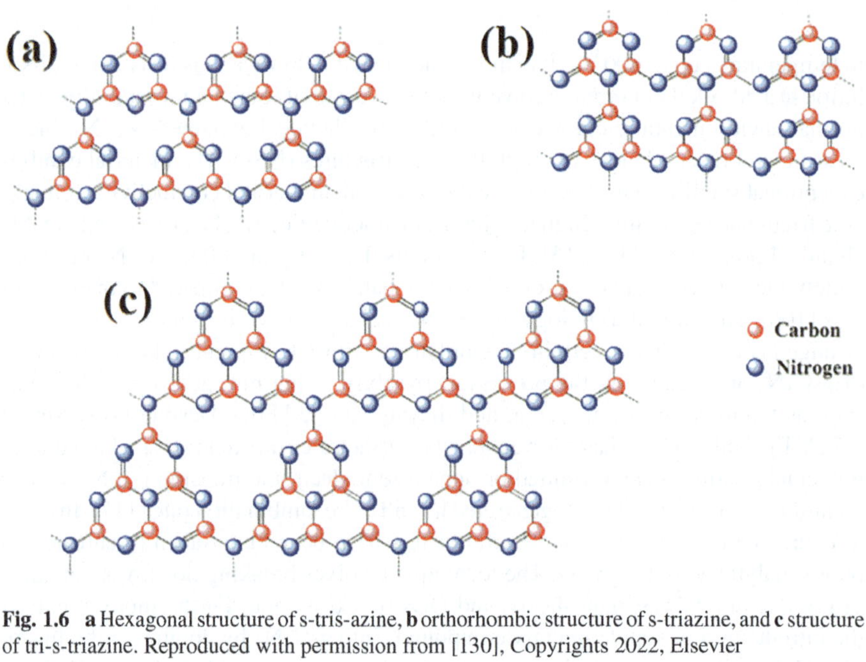


**Fig. 1.6** **a** Hexagonal structure of s-tris-azine, **b** orthorhombic structure of s-triazine, and **c** structure of tri-s-triazine. Reproduced with permission from [130], Copyrights 2022, Elsevier

are regarded as effective methods for introducing impurities into pure g-$C_3N_4$ in order to create a changed electrical structure and Eg [129]. So far, scientists have documented many methods to change the unmodified g-$C_3N_4$, including elemental and molecular doping, noble metal loading, integration with carbonaceous material, and coupling with SC [106].

## References


1. Tian, P., Tang, L., Teng, K.S., Lau, S.P.: Graphene quantum dots from chemistry to applications. Mater. Today Chem. **10**, 221–258 (2018)
2. M. I, F.Y. K.: Carbon materials science and engineering: from fundamentals to applications. 清华大学出版社 8–16 (2006)
3. Brodie, B.C.: XIII. On the atomic weight of graphite. Philos. Trans. R Soc. Lond. **149**, 249–259 (1859). https://doi.org/10.1098/rstl.1859.0013
4. Kohlschütter, V., Haenni, P.: Zur Kenntnis des Graphitischen Kohlenstoffs und der Graphitsäure. Zeitschrift für Anorg und Allg Chemie **105**, 121–144 (1919). https://doi.org/10.1002/zaac.19191050109
5. Bernal, J.D.: The structure of graphite. Proc. R. Soc. Lond. Ser. A, Contain. Pap a Math. Phys. Char. **106**, 749–773 (1924). https://doi.org/10.1098/rspa.1924.0101
6. Wallace, P.R.: The band theory of graphite. Phys. Rev. **71**, 622–634 (1947). https://doi.org/10.1103/PhysRev.71.622
7. Kroto, H.W., Heath, J.R., O'Brien, S.C., et al.: C60: Buckminsterfullerene. Nature (1985)
8. Hamwi, A., Mouras, S., Djurado, D., Cousseins, J.C.: New synthesis of first stage graphite intercalation compounds with fluorides. J. Fluor. Chem. **35**, 151 (1987). https://doi.org/10.1016/0022-1139(87)95120-7
9. Iijima, S.: Helical microtubules of graphitic carbon. Nature **354**, 56–58 (1991). https://doi.org/10.1038/354056a0
10. Xu, X., Ray, R., Gu, Y., et al.: Electrophoretic analysis and purification of fluorescent single-walled carbon nanotube fragments. J. Am. Chem. Soc. **126**, 12736–12737 (2004). https://doi.org/10.1021/ja040082h
11. Ponomarenko, L.A., Schedin, F., Katsnelson, M.I., et al.: Chaotic dirac billiard in graphene quantum dots. Science **320**(80-), 356–358 (2008). https://doi.org/10.1126/science.1154663
12. Wang, D., Chen, J.F., Dai, L.: Recent advances in graphene quantum dots for fluorescence bioimaging from cells through tissues to animals. Part. Part. Syst. Charact. **32**, 515–523 (2015). https://doi.org/10.1002/ppsc.201400219
13. Pan, D., Zhang, J., Li, Z., Wu, M.: Hydrothermal route for cutting graphene sheets into blue-luminescent graphene quantum dots. Adv. Mater. **22**, 734–738 (2010). https://doi.org/10.1002/adma.200902825
14. Harish, V., Tewari, D., Gaur, M. et al.: Review on nanoparticles and nanostructured materials: bioimaging, biosensing, drug delivery, tissue engineering, antimicrobial, and agro-food applications. Nanomaterials **12** (2022)
15. Barhoum, A., García-Betancourt, M.L., Jeevanandam, J., Hussien, E.A., Mekkawy, S.A., et al.: Review on natural, incidental, bioinspired, and engineered nanomaterials: history, definitions, classifications, synthesis, properties, market, toxicities, risks, and regulations. Nanomaterials **12**(2), 177 (2022)
16. Xu, P.T., Yang, J.X., Wang, K.S., et al.: Porous graphene: properties, preparation, and potential applications. Chinese Sci. Bull. **57**, 2948–2955 (2012)
17. Son, Y.W., Cohen, M.L., Louie, S.G.: Energy gaps in graphene nanoribbons. Phys. Rev. Lett. **97** (2006). https://doi.org/10.1103/PhysRevLett.97.216803

18. Bera, B.: A review on polymer, graphene and carbon nanotube: properties, synthesis and applications. Imp. J. Interdiscip. Res. (IJIR) **3**, 61–70 (2017)
19. Sharon, M., Sharon, M.: Graphene manufacturing applications: an introduction to the fundamentals and industrial uses (2015)
20. Rao, C.N.R., Biswas, K., Subrahmanyam, K.S., Govindaraj, A.: Graphene, the new nanocarbon. J. Mater. Chem. **19**, 2457–2469 (2009). https://doi.org/10.1039/b815239j
21. Choi, Y.S., Yoo, J.M., Hong, B.H.: Structure and properties of graphene. In: Graphene for Flexible Lighting and Displays, pp. 5–26 (2019)
22. Bhuyan, M.S.A., Uddin, M.N., Islam, M.M., et al.: Synthesis of graphene. Int. Nano Lett. **6**, 65–83 (2016)
23. Mbayachi, V.B., Ndayiragije, E., Sammani, T., et al.: Graphene synthesis, characterization and its applications: a review. Res. Chem. **3**, 100163 (2021)
24. Narita, A., Feng, X., Müllen, K.: Bottom-up synthesis of chemically precise graphene nanoribbons. Chem. Rec. **15**, 295–309 (2015). https://doi.org/10.1002/tcr.201402082
25. Haque, E., Kim, J., Malgras, V., et al.: Recent advances in graphene quantum dots: synthesis, properties, and applications. Small Methods **2**, 1–14 (2018)
26. Younis, M.R., He, G., Lin, J., Huang, P.: Recent advances on graphene quantum dots for bioimaging applications. Front. Chem. **8** (2020)
27. Molaei, M.J.: A review on nanostructured carbon quantum dots and their applications in biotechnology, sensors, and chemiluminescence. Talanta **196**, 456–478 (2019)
28. Smith, A.T., LaChance, A.M., Zeng, S., et al.: Synthesis, properties, and applications of graphene oxide/reduced graphene oxide and their nanocomposites. Nano Mater. Sci. **1**, 31–47 (2019). https://doi.org/10.1016/j.nanoms.2019.02.004
29. Zhu, Y., Murali, S., Cai, W., et al.: Graphene and graphene oxide: synthesis, properties, and applications. Adv. Mater. **22**, 3906–3924 (2010). https://doi.org/10.1002/adma.201001068
30. Tarcan, R., Todor-Boer, O., Petrovai, I., et al.: Reduced graphene oxide today. J. Mater. Chem. C **8**, 1198–1224 (2020)
31. Mbayachi, V., Ndayiragije, E., Sammani, T., et al.: Graphene Synthesis, Characterization and Its Applications: A Review. Elsevier (2021)
32. Fischer, F.R.: Bottom-up synthesis of graphene nanoribbons on surfaces. In: Advances in Polymer Science, pp. 33–65 (2017)
33. Chung, S., Revia, R.A., Zhang, M.: Graphene quantum dots and their applications in bioimaging, biosensing, and therapy. Adv. Mater. **33**, 1904362 (2021)
34. Mmaduka Obodo, R., Ahmad, I., Ifeanyichukwu Ezema, F.: Introductory chapter: graphene and its applications. In: Graphene and Its Derivatives—Synthesis and Applications. IntechOpen (2019)
35. Angione, M.D., Pilolli, R., Cotrone, S., et al.: Carbon based materials for electronic biosensing. Mater. Today **14**, 424–433 (2011)
36. Thostenson, E.T., Ren, Z., Chou, T.W.: Advances in the science and technology of carbon nanotubes and their composites: a review. Compos. Sci. Technol. **61**, 1899–1912 (2001). https://doi.org/10.1016/S0266-3538(01)00094-X
37. Ebert, L.B.: Science of fullerenes and carbon nanotubes. Carbon N. Y. **35**, 437–438 (1997). https://doi.org/10.1016/s0008-6223(97)89618-2
38. Yakobson, B.I., Samsonidze, G., Samsonidze, G.G.: Atomistic theory of mechanical relaxation in fullerene nanotubes. Carbon N. Y. **38**, 1675–1680 (2000). https://doi.org/10.1016/S0008-6223(00)00093-2
39. Yakobson, B.I., Brabec, C.J., Bernholc, J.: Nanomechanics of carbon tubes: instabilities beyond linear response. Phys. Rev. Lett. **76**, 2511–2514 (1996). https://doi.org/10.1103/PhysRevLett.76.2511
40. Ajayan, P.M., Schadler, L.S., Giannaris, C., Rubio, A.: Single-walled carbon nanotube-polymer composites: strength and weakness. Adv. Mater. **12**, 750–753 (2000). https://doi.org/10.1002/(SICI)1521-4095(200005)12:10%3c750::AID-ADMA750%3e3.0.CO;2-6
41. Iijima, S., Ichihashi, T.: Single-shell carbon nanotubes of 1-nm diameter. Nature **363**, 603–605 (1993). https://doi.org/10.1038/363603a0

42. Vamvakaki, V., Tsagaraki, K., Chaniotakis, N.: Carbon nanofiber-based glucose biosensor. Anal. Chem. **78**, 5538–5542 (2006). https://doi.org/10.1021/ac060551t
43. Grunlan, J.C., Liu, L., Kim, Y.S.: Tunable single-walled carbon nanotube microstructure in the liquid and solid states using poly(acrylic acid). Nano Lett. **6**, 911–915 (2006). https://doi.org/10.1021/nl052486t
44. Zhang, L., Li, Y., Yu, J.C.: Chemical modification of inorganic nanostructures for targeted and controlled drug delivery in cancer treatment. J. Mater. Chem. B **2**, 452–470 (2014). https://doi.org/10.1039/c3tb21196g
45. Sapsford, K.E., Algar, W.R., Berti, L., et al.: Functionalizing nanoparticles with biological molecules: developing chemistries that facilitate nanotechnology. Chem. Rev. **113**, 1904–2074 (2013)
46. Liu, Z., Sun, X., Nakayama-Ratchford, N., Dai, H.: Supramolecular chemistry on water-soluble carbon nanotubes for drug loading and delivery. ACS Nano **1**, 50–56 (2007). https://doi.org/10.1021/nn700040t
47. Dayyoub, T., Maksimkin, A.V., Filippova, O.V., et al.: Shape memory polymers as smart materials: a review. Polymers (Basel). **14** (2022)
48. Lee, H.F., Yu, H.H.: Study of electroactive shape memory polyurethane-carbon nanotube hybrids. Soft Matter. **7**, 3801–3807 (2011). https://doi.org/10.1039/c0sm01101k
49. You, Y.Z., Yan, J.J., Yu, Z.Q., et al.: Multi-responsive carbon nanotube gel prepared via ultrasound-induced assembly. J. Mater. Chem. **19**, 7656–7660 (2009). https://doi.org/10.1039/b910460g
50. Rafique, I., Kausar, A., Anwar, Z., Muhammad, B.: Exploration of epoxy resins, hardening systems, and epoxy/carbon nanotube composite designed for high performance materials: a review. Polym. Plast. Technol. Eng. **55**, 312–333 (2016). https://doi.org/10.1080/03602559.2015.1070874
51. Wang, S., Bai, J., Innocent, M.T., et al.: Lignin-based carbon fibers: formation, modification and potential applications. Green Energy Environ. **7**, 578–605 (2022)
52. Baker, D.A., Rials, T.G.: Recent advances in low-cost carbon fiber manufacture from lignin. J. Appl. Polym. Sci. **130**, 713–728 (2013)
53. Xu, Y., Liu, Y., Chen, S., Ni, Y.: Current overview of carbon fiber: toward green sustainable raw materials. BioResources **15**, 7234–7259 (2020). https://doi.org/10.15376/BIORES.15.3.XU
54. Mainka, H., Täger, O., Körner, E., et al.: Lignin: an alternative precursor for sustainable and cost-effective automotive carbon fiber. J. Mater. Res. Technol. **4**, 283–296 (2015). https://doi.org/10.1016/j.jmrt.2015.03.004
55. Edie, D.D.: Pitch and mesophase fibers. In: Carbon Fibers Filaments and Composites, pp. 43–72. Springer, Netherlands (1990)
56. Karacan, I., Erdogan, G.: The influence of thermal stabilization stage on the molecular structure of polyacrylonitrile fibers prior to the carbonization stage. Fibers Polym. **13**, 295–302 (2012). https://doi.org/10.1007/s12221-012-0295-5
57. Mahato, K.K., Dutta, K., Ray, B.C.: Static and dynamic behavior of fibrous polymeric composite materials at different environmental conditions. J. Polym. Environ. **26**, 1024–1050 (2018). https://doi.org/10.1007/s10924-017-1001-x
58. Huang, X.: Fabrication and properties of carbon fibers. Materials (Basel). **2**, 2369–2403 (2009)
59. Naskar, A.K., Walker, R.A., Proulx, S., et al.: UV assisted stabilization routes for carbon fiber precursors produced from melt-processible polyacrylonitrile terpolymer. Carbon N. Y. **43**, 1065–1072 (2005). https://doi.org/10.1016/j.carbon.2004.11.047
60. Paiva, M.C., Kotasthane, P., Edie, D.D., Ogale, A.A.: UV stabilization route for melt-processible PAN-based carbon fibers. Carbon 1399–1409 (2003)
61. Godshall, D., Rangarajan, P., Baird, D.G., et al.: Incorporation of methyl acrylate in acrylonitrile based copolymers: effects on melting behavior. Polymer (Guildf) **44**, 4221–4228 (2003). https://doi.org/10.1016/S0032-3861(03)00369-0
62. Marcuzzo, J.S., Otani, C., Polidoro, H.A., Otani, S.: Influence of thermal treatment on porosity formation on carbon fiber from textile PAN. Mater. Res. **16**, 137–144 (2013). https://doi.org/10.1590/S1516-14392012005000165

63. Chae, H.G., Minus, M.L., Kumar, S.: Oriented and exfoliated single wall carbon nanotubes in polyacrylonitrile. Polymer (Guildf) **47**, 3494–3504 (2006). https://doi.org/10.1016/j.polymer.2006.03.050
64. Sawai, D., Kanamoto, T., Yamazaki, H., Hisatani, K.: Dynamic mechanical relaxations in poly(acrylonitrile) with different stereoregularities. Macromolecules **37**, 2839–2846 (2004). https://doi.org/10.1021/ma0352330
65. Sawai, D., Kanamoto, T., Porter, R.S.: Differential scanning calorimetry evidence for the existence of a first-order thermal transition in ultraoriented at-poly(acrylonitrile). Macromolecules **31**, 2010–2012 (1998). https://doi.org/10.1021/ma971248k
66. Manera, C., Fragoso, H.P., Agra, A.A., et al.: Systematic mapping of studies on coal tar and pitch over the last five decades (1970–2023). Chem. Eng. Res. Des. **196**, 427–450 (2023)
67. Barr, J.B., Chwastiak, S., Didchenko, R., et al.: High modulus carbon fibres from pitch precursor (1976)
68. Lim, T.H., Kim, M.S., Yeo, S.Y., Jeong, E.: Preparation and evaluation of isotropic and mesophase pitch-based carbon fibers using the pelletizing and continuous spinning process. J. Ind. Text. **48**, 1242–1253 (2019). https://doi.org/10.1177/1528083718763774
69. Ward, C.: Boat-building and its social context in early Egypt: interpretations from the first dynasty boat-grave cemetery at Abydos. Antiquity **80**, 118–129 (2006). https://doi.org/10.1017/S0003598X00093303
70. Apicella, B., Tregrossi, A., Stanzione, F., et al.: Analysis of petroleum and coal tar pitches as large PAH. Chem. Eng. Trans. **57**, 775–780 (2017). https://doi.org/10.3303/CET1757130
71. Chung, D.D.L.: Carbon fiber composites (2012)
72. Lin, J., Koda, K., Kubo, S., et al.: Improvement of mechanical properties of softwood lignin-based carbon fibers. J. Wood Chem. Technol. **34**, 111–121 (2014). https://doi.org/10.1080/02773813.2013.839707
73. Sudo, K., Shimizu, K., Nakashima, N., Yokoyama, A.: A new modification method of exploded lignin for the preparation of a carbon fiber precursor. J. Appl. Polym. Sci. **48**, 1485–1491 (1993). https://doi.org/10.1002/app.1993.070480817
74. Sudo, K., Shimizu, K.: A new carbon fiber from lignin. J. Appl. Polym. Sci. **44**, 127–134 (1992). https://doi.org/10.1002/app.1992.070440113
75. Uraki, Y., Kubo, S., Nigo, N., et al.: Preparation of carbon fibers from organosolv lignin obtained by aqueous acetic acid pulping. Holzforschung **49**, 343–350 (1995). https://doi.org/10.1515/hfsg.1995.49.4.343
76. Kadla, J.F., Kubo, S., Venditti, R.A., et al.: Lignin-based carbon fibers for composite fiber applications. Carbon N Y **40**, 2913–2920 (2002). https://doi.org/10.1016/S0008-6223(02)00248-8
77. Kubo, S., Kadla, J.F.: Lignin-based carbon fibers: effect of synthetic polymer blending on fiber properties. J. Polym. Environ. **13**, 97–105 (2005). https://doi.org/10.1007/s10924-005-2941-0
78. Bajpai, P.: Lignin as a raw material for carbon fibre. In: SpringerBriefs in Materials, pp. 29–32. Springer (2017)
79. Baker, D.A., Gallego, N.C., Baker, F.S.: On the characterization and spinning of an organic-purified lignin toward the manufacture of low-cost carbon fiber. J. Appl. Polym. Sci. **124**, 227–234 (2012). https://doi.org/10.1002/app.33596
80. Compere, A.L., Griffith, W.L., Leitten, C.F., Pickel, J.M.: Evaluation of lignin from alkaline-pulped hardwood black liquor (2005)
81. Frank, E., Hermanutz, F., Buchmeiser, M.R.: Carbon fibers: precursors, manufacturing, and properties. Macromol. Mater. Eng. **297**, 493–501 (2012)
82. Xia, Y., Gates, B., Yin, Y., Lu, Y.: Monodispersed colloidal spheres: old materials with new applications. Adv. Mater. **12**, 693–713 (2000). https://doi.org/10.1002/(SICI)1521-4095(200005)12:10%3c693::AID-ADMA693%3e3.0.CO;2-J
83. Serp, P., Feurer, R., Kalck, P., et al.: Chemical vapour deposition process for the production of carbon nanospheres. Carbon N. Y. **39**, 621–626 (2001)

84. Buchholz, D.B., Doherty, S.P., Chang, R.P.H.: Mechanism for the growth of multiwalled carbon-nanotubes from carbon black. Carbon N. Y. **41**, 1625–1634 (2003). https://doi.org/10.1016/S0008-6223(03)00110-6
85. Studart, A.R., Shum, H.C., Weitz, D.A.: Arrested coalescence of particle-coated droplets into nonspherical supracolloidal structures. J. Phys. Chem. B **113**, 3914–3919 (2009). https://doi.org/10.1021/jp806795c
86. Wissler, M.: Graphite and carbon powders for electrochemical applications. J. Power. Sources **156**, 142–150 (2006)
87. Pol, V.G., Pol, S.V., Calderon-Moreno, J.M., et al.: The dependence of the oriented growth of carbon filaments on the intensity of a magnetic field. Carbon N. Y. **44**, 1913–1918 (2006). https://doi.org/10.1016/j.carbon.2006.02.007
88. Pol, V.G., Pol, S.V., Gedanken, A., et al.: Magnetic field guided formation of long carbon filaments (sausages). Carbon N. Y. **42**, 2738–2741 (2004). https://doi.org/10.1016/j.carbon.2004.05.016
89. Ding, S.J., Zhang, C.L., Yang, M., et al.: Template synthesis of composite hollow spheres using sulfonated polystyrene hollow spheres. Polymer (Guildf) **47**, 8360–8366 (2006). https://doi.org/10.1016/j.polymer.2006.10.001
90. Wu, C., Zhang, X., Ning, B., et al.: Shape evolution of new-phased lepidocrocite vooh from single-shelled to double-shelled hollow nanospheres on the basis of programmed reaction-temperature strategy. Inorg. Chem. **48**, 6044–6054 (2009). https://doi.org/10.1021/ic900416v
91. Su, F., Zhao, X.S., Wang, Y., et al.: Hollow carbon spheres with a controllable shell structure. J. Mater. Chem. **16**, 4413–4419 (2006). https://doi.org/10.1039/b609971h
92. Fang, B., Kim, J.H., Lee, C., Yu, J.S.: Hollow macroporous core/mesoporous shell carbon with a tailored structure as a cathode electrocatalyst support for proton exchange membrane fuel cells. J. Phys. Chem. C **112**, 639–645 (2008). https://doi.org/10.1021/jp710193s
93. Zhou, Z., Yan, Q., Su, F., Zhao, X.S.: Replicating novel carbon nanostructures with 3D macroporous silica template. J. Mater. Chem. **15**, 2569–2574 (2005). https://doi.org/10.1039/b503691g
94. Arnal, P.M., Weidenthaler, C., Schüth, F.: Highly monodisperse zirconia-coated silica spheres and zirconia/silica hollow spheres with remarkable textural properties. Chem. Mater. **18**, 2733–2739 (2006). https://doi.org/10.1021/cm052580a
95. Terrones, M., Grobert, N., Terrones, H.: Synthetic routes to nanoscale BxCyNz architectures. Carbon N Y **40**, 1665–1684 (2002). https://doi.org/10.1016/S0008-6223(02)00008-8
96. Winter, M., Moeller, K.-C., Besenhard, J.O.: Carbonaceous and graphitic anodes. In: Lithium Batteries, pp. 145–194. Springer, US (2009)
97. Hou, P.X., Orikasa, H., Yamazaki, T., et al.: Synthesis of nitrogen-containing microporous carbon with a highly ordered structure and effect of nitrogen doping on $H_2O$ adsorption. Chem. Mater. **17**, 5187–5193 (2005). https://doi.org/10.1021/cm051094k
98. Yang, Q.H., Hou, P.X., Unno, M., et al.: Dual Raman features of double coaxial carbon nanotubes with N-doped and B-doped multiwalls. Nano Lett. **5**, 2465–2469 (2005). https://doi.org/10.1021/nl051779j
99. Wang, W., Xu, D., Cheng, B., et al.: Hybrid carbon@$TiO_2$ hollow spheres with enhanced photocatalytic $CO_2$ reduction activity. J. Mater. Chem. A **5**, 5020–5029 (2017). https://doi.org/10.1039/c6ta11121a
100. Kim, M., Xin, R., Earnshaw, J., et al.: MOF-derived nanoporous carbons with diverse tunable nanoarchitectures. Nat. Protoc. **17**, 2990–3027 (2022). https://doi.org/10.1038/s41596-022-00718-2
101. Kim, M., Firestein, K.L., Fernando, J.F.S., et al.: Strategic design of Fe and N co-doped hierarchically porous carbon as superior ORR catalyst: from the perspective of nanoarchitectonics. Chem. Sci. **13**, 10836–10845 (2022). https://doi.org/10.1039/d2sc02726g
102. Kapri, S., Maiti, S., Bhattacharyya, S.: Lemon grass derived porous carbon nanospheres functionalized for controlled and targeted drug delivery. Carbon N. Y. **100**, 223–235 (2016). https://doi.org/10.1016/j.carbon.2016.01.017

103. Serhan, M., Sprowls, M., Jackemeyer, D., et al.: Total iron measurement in human serum with a smartphone. In: AIChE Annual Meeting, Conference Proceedings, pp. 1–3 (2019)
104. Zhao, C., Kang, J., Li, Y., et al.: Carbon-based stimuli-responsive nanomaterials: classification and application. Cyborg. Bionic. Syst. **4** (2023)
105. Deshmukh, A.A., Mhlanga, S.D., Coville, N.J.: Carbon spheres. Mater. Sci. Eng. R Reports **70**, 1–28 (2010)
106. Vaya, D., Kaushik, B., Surolia, P.K.: Recent advances in graphitic carbon nitride semiconductor: structure, synthesis and applications. Mater. Sci. Semicond. Process. **137** (2022)
107. Wang, X., Maeda, K., Thomas, A., et al.: A metal-free polymeric photocatalyst for hydrogen production from water under visible light. Nat. Mater. **8**, 76–80 (2009). https://doi.org/10.1038/nmat2317
108. Liu, A.Y., Cohen, M.L.: Prediction of new low compressibility solids. Science **245**(80-), 841–842 (1989). https://doi.org/10.1126/science.245.4920.841
109. Corkill, J.L., Cohen, M.L.: Calculated quasiparticle band gap of 2–$C_3N_4$. Phys. Rev. B **48**, 17622–17624 (1993). https://doi.org/10.1103/PhysRevB.48.17622
110. Hemani, G.: Study of graphene growth using (2013)
111. Teter, D.M., Hemley, R.J.: Low-compressibility carbon nitrides. Science **271**(80-), 53–55 (1996). https://doi.org/10.1126/science.271.5245.53
112. Dong, G., Zhang, Y., Pan, Q., Qiu, J.: A fantastic graphitic carbon nitride (g-$C_3N_4$) material: electronic structure, photocatalytic and photoelectronic properties. J. Photochem. Photobiol. C Photochem. Rev. **20**, 33–50 (2014)
113. Kroke, E., Schwarz, M., Horath-Bordon, E., et al.: Tri-s-triazine derivatives. Part I. From trichloro-tri-s-triazine to graphitic $C_3N_4$ structures. New J. Chem. **26**, 508–512 (2002). https://doi.org/10.1039/b111062b
114. Wang, X., Blechert, S., Antonietti, M.: Polymeric graphitic carbon nitride for heterogeneous photocatalysis. ACS Catal. **2**, 1596–1606 (2012)
115. Patnaik, S., Martha, S., Parida, K.M.: An overview of the structural, textural and morphological modulations of g-$C_3N_4$ towards photocatalytic hydrogen production. RSC Adv. **6**, 46929–46951 (2016)
116. Mo, Z., Xu, H., Chen, Z., et al.: Self-assembled synthesis of defect-engineered graphitic carbon nitride nanotubes for efficient conversion of solar energy. Appl. Catal. B Environ. **225**, 154–161 (2018). https://doi.org/10.1016/j.apcatb.2017.11.041
117. Chen, S., Hao, N., Jiang, D., et al.: Graphitic carbon nitride quantum dots in situ coupling to $Bi_2MoO_6$ nanohybrids with enhanced charge transfer performance and photoelectrochemical detection of copper ion. J. Electroanal. Chem. **787**, 66–71 (2017). https://doi.org/10.1016/j.jelechem.2017.01.042
118. Deng, Y.: Insight into highly efficient simultaneous photocatalytic removal of Cr (VI) and 2,4-diclorophenol under visible light irradiation by phosphorus doped porous. Appl. Catal. B, Environ. (2016)
119. Mao, Z., Chen, J., Yang, Y., et al.: Novel g-$C_3N_4$/CoO nanocomposites with significantly enhanced visible-light photocatalytic activity for $H_2$ evolution. ACS Appl. Mater. Interfaces **9**, 12427–12435 (2017). https://doi.org/10.1021/acsami.7b00370
120. Xia, P., Zhu, B., Cheng, B., et al.: 2D/2D g-$C_3N_4$/$MnO_2$ nanocomposite as a direct Z-scheme photocatalyst for enhanced photocatalytic activity. ACS Sustain. Chem. Eng. **6**, 965–973 (2018). https://doi.org/10.1021/acssuschemeng.7b03289
121. Zhang, X., Xie, X., Wang, H., et al.: Enhanced photoresponsive ultrathin graphitic-phase $C_3N_4$ nanosheets for bioimaging. J. Am. Chem. Soc. **135**, 18–21 (2013). https://doi.org/10.1021/ja308249k
122. Du, X., Zou, G., Wang, Z., Wang, X.: A scalable chemical route to soluble acidified graphitic carbon nitride: an ideal precursor for isolated ultrathin g-$C_3N_4$ nanosheets. Nanoscale **7**, 8701–8706 (2015). https://doi.org/10.1039/c5nr00665a
123. Ye, S., Wang, R., Wu, M.Z., Yuan, Y.P.: A review on g-$C_3N_4$ for photocatalytic water splitting and $CO_2$ reduction. In: Applied Surface Science, pp. 15–27. Elsevier B.V., (2015)

124. Yang, S., Gong, Y., Zhang, J., et al.: Exfoliated graphitic carbon nitride nanosheets as efficient catalysts for hydrogen evolution under visible light. Adv. Mater. **25**, 2452–2456 (2013). https://doi.org/10.1002/adma.201204453
125. Rono, N., Kibet, J.K., Martincigh, B.S., Nyamori, V.O.: A review of the current status of graphitic carbon nitride. Crit. Rev. Solid State Mater. Sci. **46**, 189–217 (2021)
126. Anjumol, K., Joy, J., Mathew, S., et al.: Graphitic Carbon Nitride-Based Nanocomposites. Elsevier
127. Ye, C., Li, J.X., Li, Z.J., et al.: Enhanced driving force and charge separation efficiency of protonated g-$C_3N_4$ for photocatalytic $O_2$ evolution. ACS Catal. **5**, 6973–6979 (2015). https://doi.org/10.1021/acscatal.5b02185
128. Liu, H., Wang, X., Wang, H., Nie, R.: Synthesis and biomedical applications of graphitic carbon nitride quantum dots. J. Mater. Chem. B **7**, 5432–5448 (2019)
129. Gao, J., Wang, Y., Zhou, S., et al.: A facile one-step synthesis of Fe-doped g-$C_3N_4$ nanosheets and their improved visible-light photocatalytic performance. Chem. Cat. Chem. **9**, 1708–1715 (2017). https://doi.org/10.1002/cctc.201700492
130. Vaya, D., Kaushik, B., Surolia, P.K.: Recent advances in graphitic carbon nitride semiconductor: structure, synthesis and applications. Mater. Sci. Semicond. Process. **137**, 106181 (2022)

# Chapter 2
# Synthesis Protocol

## 2.1 Synthesis of Graphene

### 2.1.1 Mechanical Exfoliation

Mechanical exfoliation is a highly unusual and distinguished method used to remove individual layers of Gr flakes onto desired surfaces. This is the initial acknowledged technique for synthesizing Gr. This nanotechnology method applies a stress, either longitudinal or transverse, to the surface of layered structure materials [1]. Geim and Novoselov initially pioneered this technique, employing Scotch tape to iteratively exfoliate highly oriented pyrolytic Gt, resulting in the production of progressively thinner sheets. A commonly used technique involves using tape to separate the layers of Gt and then transferring the thinner layers of Gr onto a different piece of tape (Fig. 2.1a–g) [2]. Gt is formed by the arrangement of individual layers of monoatomic Gr, which are bonded together by relatively weak van der Waals interactions. The separation between layers and the energy needed to rupture the interlayer bonds are 3.34 Å and 2 eV/nm$^2$, correspondingly. To separate one mono-atomic layer from Gt, mechanical cleaving is necessary, which involves applying an external force of around 300 nN/$\mu$m$^2$. Sheet stacking in Gt is caused by partly filled $\pi$ orbitals overlapping transverse to the sheet plane. The occurrence of this overlap is made possible by the presence of van der Waals forces. Exfoliation is the process that happens when larger lattice spacing and weaker bonding in the transverse orientation cause the opposite of stacking, compared to the robust bonding and smaller lattice spacing in the hexagonal structure [3]. Gr sheets of various dimensions can be generated by mechanical exfoliation or layer removal from graphitic materials like single-crystal Gt, highly ordered pyrolytic graphite (HOPG), or natural Gt [4–7]. Exfoliation can be accomplished using several techniques, including the ultrasonication, electric field, scotch tape, and transfer printing technologies. Some studies have utilized conventional adhesives like epoxy resin or self-assembled monolayers (SAMs) to attach HOPG to a substrate. This is done to enhance the production of



M. Ikram et al., *Carbon-Based Nanomaterials for Environmental Applications*,
Engineering Materials, https://doi.org/10.1007/978-3-031-59390-1_2

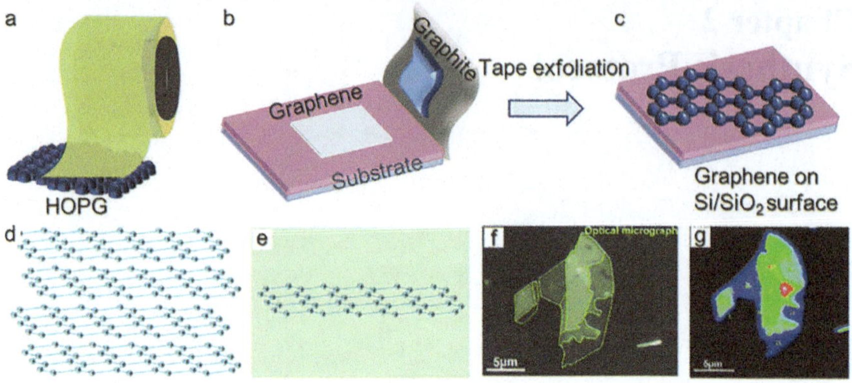


**Fig. 2.1** Mechanical exfoliation of Gr **a** HOPG layers are separated by repeatedly sticking tape; **b** Si/$SiO_2$ substrate obtains the layered sheet; **c** after progressively peeling the tape, Gr is on the substrate; **d** diagram of few-layer Gr; **e** monolayer Gr; **f** optical micrograph; and **g** exfoliated Gr four-wave mixing micrograph. Reproduced with permission from [2], Copyright 2021, Wiley

single and few-layer Gr flakes [8]. Furthermore, a recent work showcases the process of transferring large-scale Gr designs from patterned HOPG utilizing gold sheets [9]. Producing high-quality Gr using this technology is far more cost-effective than any other method. The micromechanical exfoliation process involves the separation of Gr from a Gt crystal by use of sticky tape. Upon removing it from the Gt, the tape retains multiple-layer Gr. Through iterative exfoliation, the multiple-layer Gr is fragmented into numerous flakes consisting of only a few layers of Gr. Subsequently, the tape is adhered to the acetone substrate to facilitate its detachment. Finally, a last peeling is performed using a leftover adhesive tape. The resultant flakes display notable disparities in both dimensions and thickness, with diameters spanning from nanometers to many tens of micrometers for individual layers of Gr, dependent on the wafer employed. Although single-layer graphene has a low absorption rate of 2%, it may nevertheless be observed under a light microscope on $SiO_2$/Si owing to the interference effect [10].

### *2.1.2 Electrochemical Exfoliation*

Electrochemical exfoliation of Gr is a viable technique for the extensive production of Gr from Gt, serving as an alternative to mechanical or oxidation-driven approaches [11–13]. When a voltage is applied, the ionic species in an electrolyte are attracted to and enter the Gt electrode, resulting in an increase in the distance between the layers [14]. In ammonium sulfate, sulfate ions and water molecules generate $SO_2$ and $O_2$ gas bubbles between Gt layers. This causes the neighboring layers to separate from one other, as seen in Fig. 2.2a [15]. Notwithstanding these advancements in electrochemical exfoliation, there exist many significant obstacles confronting the field: (i) Electrochemical exfoliation requires the usage of Gt monoliths instead of loose Gt powders. This is because the Gt electrode needs to be uninterrupted, capable of conducting electricity, and linked to the external power supply. (ii) Crucially, if electrochemical exfoliation indeed takes place and causes the monolith to grow into Gr, the monolith will collapse, therefore halting the electrochemical exfoliation process. As intercalation progresses, the Gt electrode has a tendency to disintegrate. When this happens, the Gt fragments get detached from the monolith and are no longer able to take part in the intercalation process. Deterioration of the Gt rod (Fig. 2.2b) limits Gr production and requires additional separation of unexfoliated compounds. Liu et al. placed the monolithic Gt electrode at the base of the exfoliation vessel to use gravity to keep fragmented Gt particles together [12]. Using sonicated dispersions in dimethyl formamide, yield increased somewhat, but graphitic particle retention is uncontrolled. Abdelkader et al. suggested a comparable arrangement in which a constant stream of electrolyte solution is continually supplied to the setup, enabling the collection of any detached, suspended Gr [12]. Wang et al. showed that using a paraffin layer on the Gt electrode may successfully restrict the electrochemical exfoliation to a certain volume [16]. Nevertheless, in this particular instance, post-processing involved the utilization of intense sonication in *N*-methyl-2-pyrrolidone, followed by centrifugation. Figure 2.2c demonstrates the use of Gt flakes, instead of a solid Gt rod electrode, by Achee et al. These flakes were placed inside a permeable container and served as the working electrode, without the need for any binder [17].

### *2.1.3 Preparation of Gr by Chemical Oxidation–Reduction*

The process of oxidatively exfoliating Gt is widely recognized as the most viable technique for generating GtO and the associated compound, GO. GO is frequently employed as a primary raw material for the dense production of Gr products. This technique entails the incorporation of oxygen functional groups by oxidation, which then distribute evenly within the C framework to counteract the inter-sheet van der Waals forces. The presence of oxygen functions leads to an expansion of the distance between the layers of Gt sheets. The structure and chemical composition of GtO are mainly influenced by the Gt precursor, oxidizing agents, synthesis process, and reaction circumstances. GtO is a non-stoichiometric material. By subjecting GtO to sonication in appropriate solvents such as ethanol or water, it may be transformed into uniformly distributed single-, bi-, and few-layer GO [18–20]. Moreover, GO may be converted into rGO using many reduction techniques, including thermal, chemical, electrochemical, hydrothermal, microwave, and photoreduction procedures (Fig. 2.3). The process of chemically reducing GtO is a commonly used method for producing significant quantities of Gr [21]. GtO is often produced by oxidizing Gt with strong sulfuric acid, nitric acid, or potassium permanganate, utilizing methods such as the Brodie process, Staudenmaier technique, or Hummers method [22–24]. An alternative method for producing graphene involves the use of sonication and

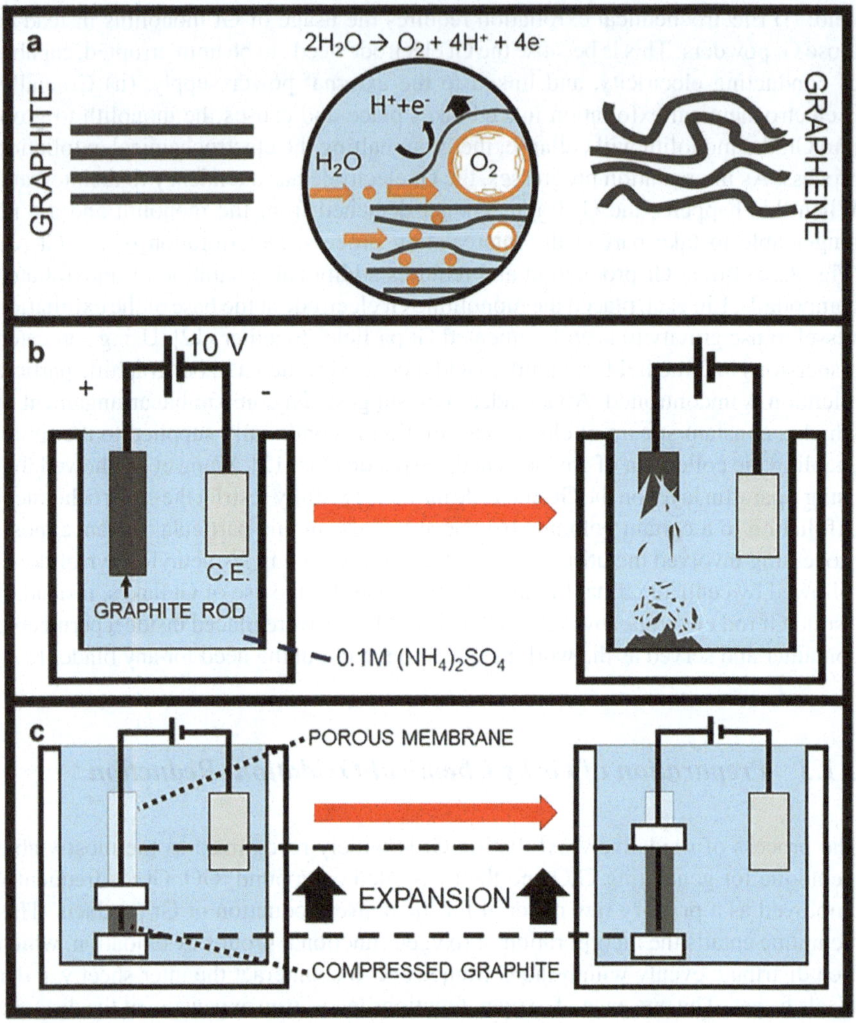


**Fig. 2.2** **a** Electrochemical exfoliation of Gr: transformation of Gt to Gr as sulfate intercalation and oxidation proceeds. **b** Previous attempts to employ this notion inevitably disintegrated the Gt working electrode after intercalation. **c** Electrochemical Gt flakes exfoliation in a porous, extensible container. Reproduced with permission from [17], Copyright 2018, Springer Nature

reduction of GO. Hydrogen gas ($H_2$) is added to the alkenes, resulting in a simultaneous discharge of nitrogen gas. An excessive amount of $NaBH_4$ has been used as a reductant [25]. The other reducing agents used are hydroxylamine, ascorbic acid, alkaline solutions, glucose, hydroquinone, phenyl hydrazine, and pyrrole [26–30]. GO was formed by the chemical reaction among organic isocyanates and OH groups.

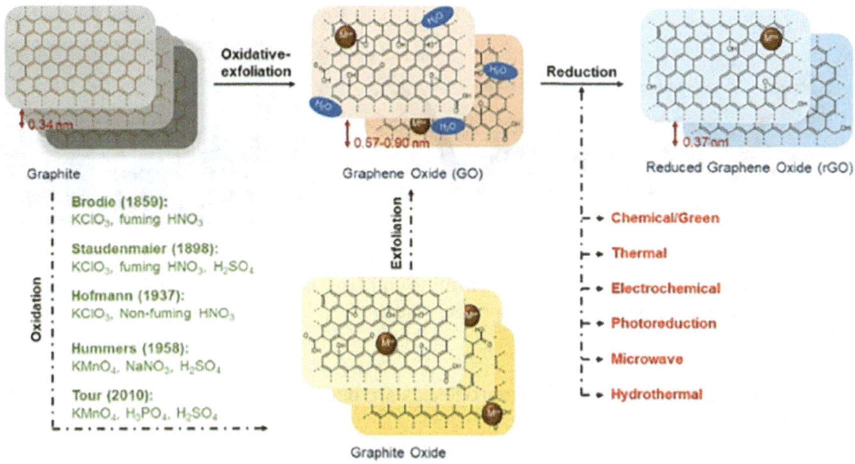


**Fig. 2.3** Chemical oxidation–reduction method used to prepare GO and rGO from Gt. Reproduced with permission from [31], Copyright 2021, Elsevier

## *2.1.4 Chemical Vapor Deposition Method*

CVD is a commonly used technique for producing CBNMs due to its simplicity and effective control capabilities [32–35]. Somani et al. presented the inaugural research in 2006 on the utilization of solid transition metal catalysts for the synthesis of Gr [36]. Despite addressing the scalability issues encountered in mechanical exfoliation, the CVD technique is expensive due to the need for advanced tools, high energy consumption, and the time-consuming and costly process of transferring Gr. Researchers are now focused on discovering answers to these difficulties. Many researchers are primarily focused on minimizing the expense associated with the production of Gr via CVD. Several collaborative endeavors have been undertaken to decrease the expense of CVD Gr, such as the advancement of energy-efficient techniques, the reuse of the substrate, and the simplification of the Gr transfer process [37]. Researchers have achieved a notable accomplishment by continuously mass-producing centimeter-wide Gr sheets that were up to around 100 m long. They used a roll-to-roll CVD technique for this purpose, as shown in Fig. 2.4 [38–40]. A recent research successfully generated uniform, larger area, high-quality single-layer Gr (ID/IG $< 0.065$) utilizing a continuous roll-to-roll Gr CVD manufacturing process. The production speed exceeded 5 cm per minute. The nanoporous atomic thin membrane, fabricated using roll-to-roll automated CVD of Gr, demonstrated size-dependent molecular transport properties comparable to conventionally constructed graphene membranes. Researchers also investigated the roll-to-roll CVD technique to produce doped Gr for the development of very lightweight solar devices [39]. Nevertheless, the Gr sheets produced by these roll-to-roll methods exhibit several limitations, including elevated sheet resistance, imperfections (such as vacancies), excessive expenses, and an excessive number of layers (multi-layer graphene or MLG).

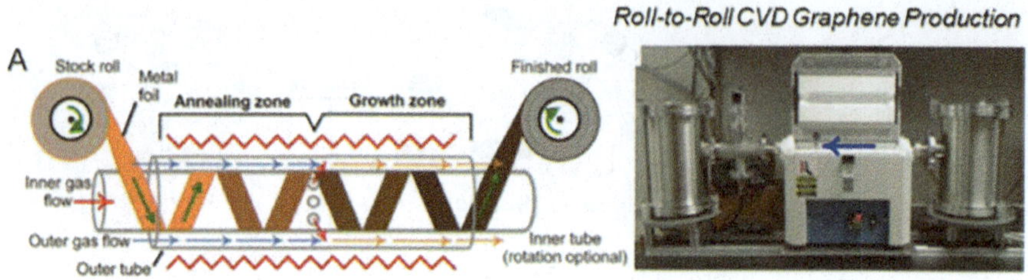


**Fig. 2.4** **a** Concentric tube CVD system for roll-to-roll Gr synthesis on Cu foil. Reproduced with permission from [40], Copyright 2015, Springer Nature

Specialized improvements of these processes are necessary to generate economically viable, high-quality Gr.

### 2.1.5 *Epitaxial Growth (SiC Heating)*

Obtaining Gr that is both high quality and covers a broad area is essential for manufacturing nanoelectronics that rely on Gr. As previously mentioned, other techniques for Gr synthesis have inherent limitations. The CVD method necessitates a purification procedure to eradicate the catalyst particles and relocate Gr to a different substrate. The chemical reduction of GO involves the use of chemicals, which can have an impact on the properties of Gr. The scotch tape technique is not perfect for massive production of Gr. By annealing silicon carbide (SiC) as an insulator at high temperatures in vacuum, epitaxial Gr on silicon carbide is growing rapidly. This process results in the formation of a graphitic surface by sublimating Si atoms [41, 42] (refer to Fig. 2.5a). The significant promise of epitaxial Gr as a novel electronic material is well acknowledged [43]. Gr was epitaxially grown on SiC substrates. The substrate is subjected to high temperatures, namely, around 1250 °C, and low pressures, approximately about 10–6 Torr. Ultrathin layers of Gt were created using the thermal decomposition process on the (0001) surface of 6H–SiC. On average, the films are composed of three Gr sheets [44]. The epitaxial development of Gr on SiC is considered an ideal material for advanced electronics, with the potential to outperform silicon in important areas such as feature size, speed, and power consumption [41]. Although the epitaxial growth approach is capable of producing high-quality Gr with exceptional characteristics, achieving a high production rate with this method is challenging.

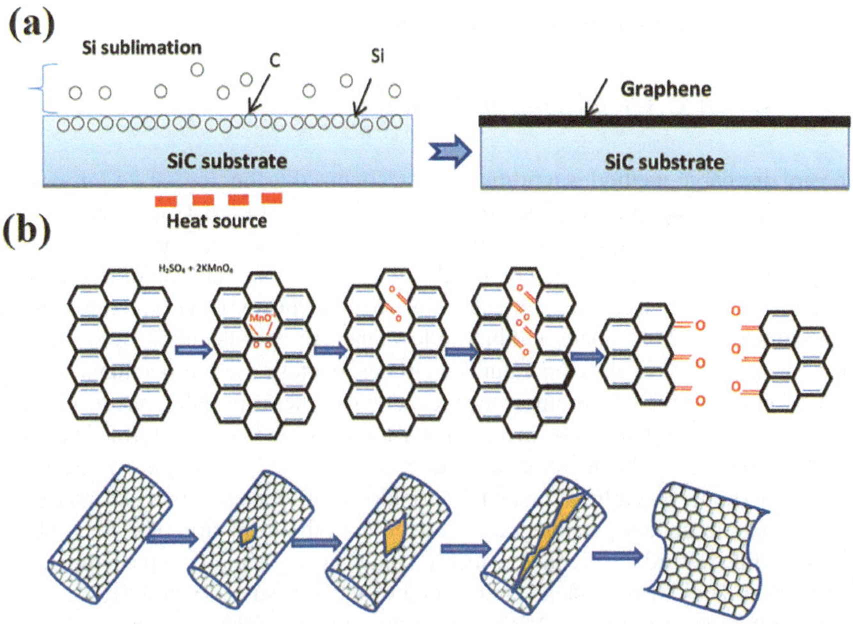


Fig. 2.5 **a** Epitaxial development of Gr, **b** unzipping of CNTs. Reproduced with permission from [48], Copyright 2021, National Library of Medicine

## 2.1.6 *Unzipping of CNTs*

Saito et al. [45] demonstrated that MWCNTs may be reduced to nanometer-sized CNTs using the process of sonication in a mixture of concentrated $H_2SO_4$ and $HNO_3$ [43]. By suspending MWCNTs in concentrated $H_2SO_4$ and then treating those with potassium permanganate ($KMnO_4$), Kosynkin et al. [46] successfully created oxidized Gr nanoribbons. The Gr nanoribbons exhibited exceptional solubility in water, ethanol, and other polar organic solvents [47]. The process of unzipping the nanotubes appears to be analogous to the unzipping of GtO, as described in references and [45, 46]. The opening can take place either longitudinally or spirally, depending on the location being targeted and the angle of chirality. The primary cause of alkene structure degradation in CNT is the generation of manganate esters ($O{=}MnO^{-2}{=}O$). This process renders the first attacked sites more susceptible to subsequent intense reactions with permanganate [44]. Figure 2.5b depicts the suggested process of unzipping CNTs.

## 2.2 Synthesis of Carbon Nanotubes

### 2.2.1 *Electric Arc Discharge Method*

The arc discharge method is a primary and recognized technology used for synthesizing CNTs. Originally, it was created for the production of $C_{60}$ fullerenes and subsequently adapted for the manufacturing of MWCNTs and SWCNTs [47, 49, 50]. The arc discharge method commonly employs the vaporization of two C electrodes by the use of an electric arc. These electrodes are positioned in close proximity, approximately 1–2 mm apart, within a hollow that is filled with either a low-pressure inert gas or liquids. High-temperature electrodes, operating at temperatures ranging from about 4000–6000 K, promote the breakdown of the surrounding media (such as gas or liquid) in the chamber by means of resistive heating. Plasma is formed between the electrodes due to the moderate ionization. The elevated temperature of the anode causes C to undergo sublimation and evaporation. The heat energy in plasma causes the C vapors to undergo disintegration, resulting in the formation of C ions. The C gases migrate toward the cathode, which is somewhat cooler, due to the temperature differential. They then condense into liquid C and subsequently solidify, resulting in the formation of rod-shaped deposits on the cathode. These deposits are formed of CNTs [51]. Nevertheless, there are still uncertainties surrounding the process of CNTs creation. There are other views on the mechanism of growth [52, 53].

The presence of catalysts is one of the elements that influence the formation of CNTs. The experimental findings have shown that the use of a metal catalyst endorses the development of SWCNTs over MWCNTs [54]. Additionally, the size of the catalyst impacts the distribution of diameters. Both Nickel (Ni) and Iron (Fe) can enhance the yield and increase the quality of SWNTs manufacturing. Thus, Gamaly and Ebbesen illustrated the use of electrodes equipped with perforated openings, containing metals such as Fe, Ni, and Co, while the chamber was filled with helium at a pressure range of 100–500 Torr (Fig. 2.6a) [54]. Although the presence of Fe contributes to the elongation of nanotubes, it also leads to the formation of $Fe_2O_3$ oxide, which complicates the creation of CNTs. Therefore, it is recommended that the concentration of Fe does not surpass 10% in order to create CNTs [55]. Nevertheless, in the context of a binary catalyst, the presence of Fe compels the creation of MWCNTs. Therefore, in their study published in, Zhao et al. proposed the incorporation of sulfur together with Fe for the creation of double-walled CNTs (DWCNT) [56]. The outcome of the discharge, including the distribution of diameters and the creation of bundles of CNTs, is contingent upon the temperature and consistency of the plasma arc. The diameter distribution of CNTs is ascertained by measuring the density of C vapors in plasma. Furthermore, the C product's structure is determined by the pressure of inert gas. At a pressure of 1000 Torr, C onions are generated, whereas reducing the pressure to 500 Torr facilitates the creation of nanotubes. Additionally, a reduction to 100 and 20 Torr leads to the formation of fullerene and amorphous Gt, respectively [57]. Moreover, it has been shown that the nature of the electric current has an impact on the process of generating CNTs. As stated by

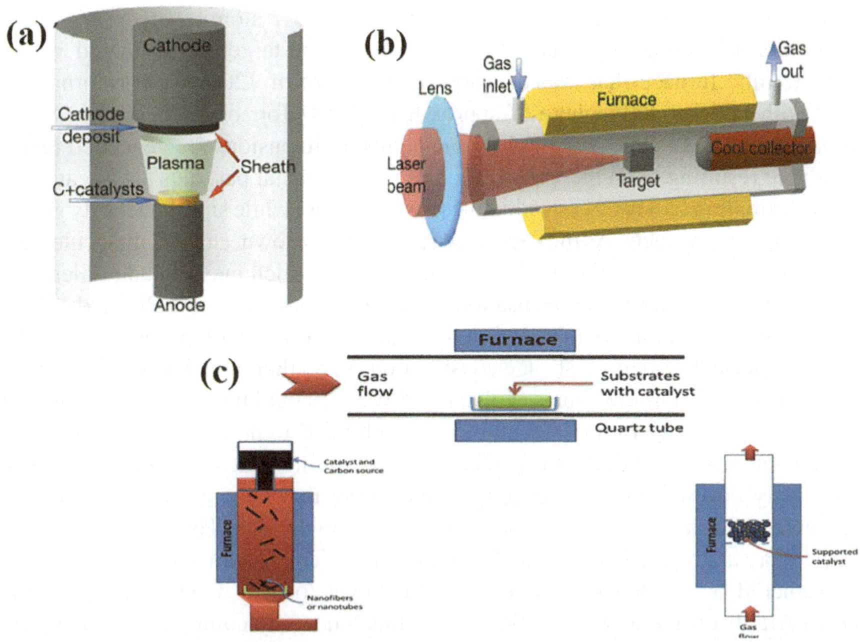


**Fig. 2.6** Schematic setup of CNT fabrication techniques: **a** arc discharge process, **b** laser ablation; Reproduced with permission from [59], Copyright 2016, Nanophotonics. Diagram of CVD method. **a** Horizontal furnace. **b** Vertical furnace. **c** Fluidized bed reactor. Reproduced with permission from [70], Copyrights 2010, National center for biotechnology information

Ashraf et al. a continuous arc discharge promotes the development of diatomic C clusters $C_2$, which is crucial for the creation of nanotubes [58]. On the other hand, a pulsed arc discharge creates monatomic clusters $C_1$. Applied DC to deposited CNTs on the cathode. However, in the case of AC deposition, the reactor chamber's sidewalls get the deposited material, as the electrodes' polarity alternates with each cycle. The short-pulse AC discharge exhibits higher energy levels in comparison to DC. Therefore, its utilization enhances the output and enables more regulation of the procedure. Ultimately, the arc discharge process stands out as a straightforward approach, enabling cost-effective manufacturing of CNTs. Nevertheless, it is plagued by the wide range of diameters in the manufactured C nanotubes, the presence of clusters, soot, and leftover catalytic particles in the end product [59].

### 2.2.2 Laser Ablation

The initial manufacture of considerable quantities (in grams) of SWCNTs was successfully accomplished in 1996 by the research team led by Smalley at Rice University [60, 61]. In this experiment, a laser, either pulsed or continuous, is employed to convert a composite target consisting of 1.2 atomic percent of cobalt/nickel and 98.8 atomic percent of Gr into vapor. The target is positioned within a quartz tube furnace that is heated to a temperature of 1200 °C. The furnace is filled with an inert atmosphere of approximately 500 Torr of either argon (Ar) or helium (He). Metal catalyst particles of nanometer dimensions are generated inside the plume resulting from the vaporization of Gt. The metal particles act as catalysts for the formation of SWCNTs within the plasma plume, while simultaneously generating many by-products. As the vaporized species cools down, small C molecules and atoms quickly come together to form larger clusters, which may include fullerenes. The catalysts also initiate condensation, albeit at a slower rate initially, and adhere to C clusters, impeding their formation into cage-like framework. Catalysts have the ability to cause the opening of cage constructions upon their attachment. From these early clusters, tubular molecules evolve into SWCNTs until the catalyst particles are too large or the temperature has dropped enough for C to no longer spread through or over the surface. Additionally, there is a possibility that the particles become sufficiently covered with a C coating to the extent that they are unable to absorb any more, resulting in the cessation of nanotube growth [61]. The SWCNTs in this scenario are aggregated due to van der Waals forces. The nanotubes and by-products are gathered by condensing them on a cold finger located downstream from the target. Arc discharge and laser ablation are fundamentally comparable techniques. Both procedures include the utilization of a metal-impregnated Gt target (anode) to generate SWCNTs. Additionally, when pure Gt is employed instead, both processes yield MWCNTs and fullerenes. Nevertheless, the length of MWCNTs generated using laser ablation is significantly shorter compared to those created using the arc discharge approach. Hence, this approach appears insufficient for the synthesis of MWNT. The diameter distribution of SWCNTs produced using this technique ranges approximately from 1.0 to 1.6 nm. Scientists are attempting to increase the scale of laser ablation due to the high quality of nanotubes created using this technique. Nevertheless, the outcomes have not reached the same level of excellence as those achieved using the arc discharge process, but they still hold great potential (Fig. 2.6b). Two recent advancements in this sector include the utilization of ultrafast pulses generated by a free electron laser and the continuous wave laser-powder approach [62].

### 2.2.3 *Chemical Vapor Deposition Method*

Although the arc discharge approach may generate a high volume of nanotubes without purification, much attention is being focused on developing production methods that provide more precise pathways for synthesizing nanotubes. CVD appears to be the most promising method for achieving a regulated procedure in the selective synthesis of nanotubes with predetermined parameters [61]. C filaments are formed by catalytic breakdown of C-containing gas on metal surfaces, a long-established process [63]. Nevertheless, the feasibility of synthesizing CNTs using this process remained unsupported until Yacamàn et al. achieved success [64].

Currently, the catalytic CVD (CCVD) method is widely regarded as the most economically feasible technique for producing CNTs on a large scale and incorporating them into different devices. CVD is essentially the process of catalytically decomposing hydrocarbon or CO feedstock using transition metal catalysts. In a flow furnace under normal atmospheric pressure, the experiment is usually done. The furnace can be horizontal or vertical (Fig. 2.6c). The use of a horizontal furnace is highly preferred. This procedure inserts a ceramic or quartz boat with the catalyst into a quartz tube. The hydrocarbon supply and inert gas are combined and passed over the catalyst bed at temperatures ranging from 500 to 1100 °C. Afterward, the system is cooled to the surrounding temperature. The vertical furnace structure is commonly used for the uninterrupted large-scale manufacturing of CNTs/fibers. The C source and catalyst are introduced into the furnace from the upper part, and the consequent filaments undergo growth while in motion and are subsequently gathered in the lower section of the chamber. The reactor can receive ultrafine metal catalyst nanoparticles either by injecting them directly or by in situ production using precursors such as metallocenes. The fluidized bed reactor is an altered iteration of the vertical furnace. The catalysts that are endorsed are usually placed in close proximity to the central part of the furnace, while the gaseous C feedstock is directed in an upward direction. During the fluidization process, the supported catalysts have a significantly longer residence time in the furnace relative to the vertical floating procedure [65–67]. The CVD approach involves the creation of CNTs by the decomposition of hydrocarbon molecules with the assistance of a transition catalyst, followed by the attachment of C onto metal nanoparticles. The metal particle undergoes C precipitation, leading to the formation of tubular C solids characterized by a $sp^2$ configuration. The characteristics of CNTs produced using the CVD method are affected by several factors. These factors include the temperature and pressure during the process, the type and pretreatment of the metallic catalyst, the concentration of hydrocarbon used, the quantity and size of the CNTs, the nature of the support material, and the duration of the reaction [68]. The regulation of CNTs diameter may be achieved by manipulating the active particles present on the catalyst's surface. The length of the tubes is determined by the duration of the reaction, and it is possible to make tubes as long as 60 mm [69].

### 2.2.4 Electrolysis

Hsu et al. discovered electrolysis as a less prevalent technique for producing CNTs in 1995 [71]. This approach primarily involves the electrowinning process of alkali metals (Li, K, Na) or alkaline-earth metals (Mg, Ca) from their chloride salts using a Gt cathode. Subsequently, the metal being deposited interacts with the cathode to generate CNTs. Distinct optimal temperatures for nanotube synthesis have been identified for each electrolyte composition, resulting in a substantial decrease in purity beyond these optimal temperatures. The optimal NaCl and LiCl electrolyte temperature was slightly over the salt's melting point. Dissolving ionic salt in purified water and sifting the liquid yields carbonaceous material after electrolysis. Erosion of the cathode during electrolysis produces CNTs and other C NPs with different morphologies. The Gt cathode produces C-encased metal particles, C filaments, and amorphous C NPs. Bai et al. grew SWCNTs instead of multi-walled CNTs [72]. The researchers synthesized CNTs by subjecting Gt to electrolytic transformation in fused NaCl at a temperature of 810 °C, with the use of argon (Ar) as an inert gas. The diameter of SWCNTs is predicted to range from 1.3 to 1.6 nm. This is similar to the SWCNTs acquired using alternative techniques. Incorporating metals or salts having a low melting point, such as $PbCl_2$, Pb, $SnCl_2$, Sn, or Bi, into the electrolyte in quantities below 1 wt% results in the formation of metal nanowires and CNTs that are filled with these metals. MWCNTs have diameters ranging from 10 to 20 nm. They are composed of a small number of walls, often 10 to 15, and their length is believed to be greater than 500 nm. MWCNTs are primarily found in tangled clusters that also include spherical C particles, amorphous C, and metal-encapsulated particles.

This method of production is unique compared to others since it takes place in the condensed phase and uses Gt as the raw material at very low temperatures. The method has several benefits: easy apparatus design, electrolysis mode control, inexpensive raw materials, low electrolysis energy consumption, and the ability to optimize electrolysis conditions and electrolytic bath composition to control product morphologies, frameworks, and C phase doping in a single step [73–75]. Novoselova et al. are now developing an innovative electrolytic synthesis technique to generate CBNMs from ionic melts [74]. The fundamental principle of the proposed technique is the electrochemical reduction of $CO_2$ to produce elemental C on metallic electrodes.

The suggested method uses cathodic reduction to convert $CO_2$ into pure C on metallic electrodes (Fig. 2.7). Electrochemical operations condense a CSs on the cathode from molten salt with dissolved carbonic acid. The electrolyte employed was a ternary combination of alkali metal chlorides, specifically NaCl, KCl, and CsCl. The primary products consisted of MWCNTs, which typically exhibited a curved shape and were frequently found in clusters or bundles. The outer tube diameter ranged from 5 to 250 nm, whereas the inner tube diameter ranged from 2 to 140 nm. The nanotubes included a minor amount of electrolyte salt. It was observed that increasing the current density results in a reduction in the diameter of CNTs at the same time, there is an increase in both the C yield and the CNTs proportion in the total mass of the product [70].

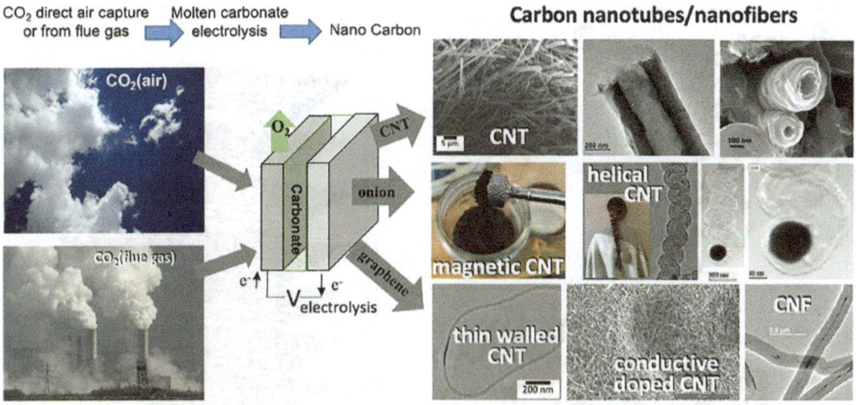


**Fig. 2.7** Electrocatalytic preparation of CNTs from $CO_2$. Reproduced with permission from [76], Copyright 2022, Catalysts

## 2.3 Synthesis of Carbon Fibers

### *2.3.1 Catalytic Thermal CVD*

Two distinct forms of CF may be synthesized using catalytic thermal deposition CVD: cup-stacked CF and platelet CF. The cup-stacked CF, often referred to as conical CF, was first identified by Ge and Sattler in 1994 [77]. The cup-shaped CF and platelet CF structures are depicted in Fig. 2.8a–d.

The CVD growth process employs several catalysts such as Fe, V, Co, Cr, and Ni, which has the capacity to dissolve C and form metal carbide, for the synthesis of CF. In addition, $CH_4$, Mo, Co, synthesis gas ($H_2$/CO), $C_2H_2$, and $C_2H_4$ are used as C precursor at 700–1200 K [78]. The topography of the catalytic NPs mostly dictates the fundamental structures of the CF. Graphitic C precipitation from hydrocarbons dispersed among metal particles proves the growth process [79].

### *2.3.2 Electrospinning*

Electrospinning is a commonly employed technique for fabricating CFs. In a recent systematic study, Inagaki thoroughly examined the preparation of CF using the electrospinning technique [81]. In a recent study, Zhang provided a comprehensive overview of the production and uses of CFs that were generated using the electrospinning technique [82].

In order to produce the CFs using the electrospinning technique, it is necessary to first manufacture the polymer nanofibers as the precursors for the CFs (Fig. 2.9). The characteristics of the resulting CF are determined by the composition of the

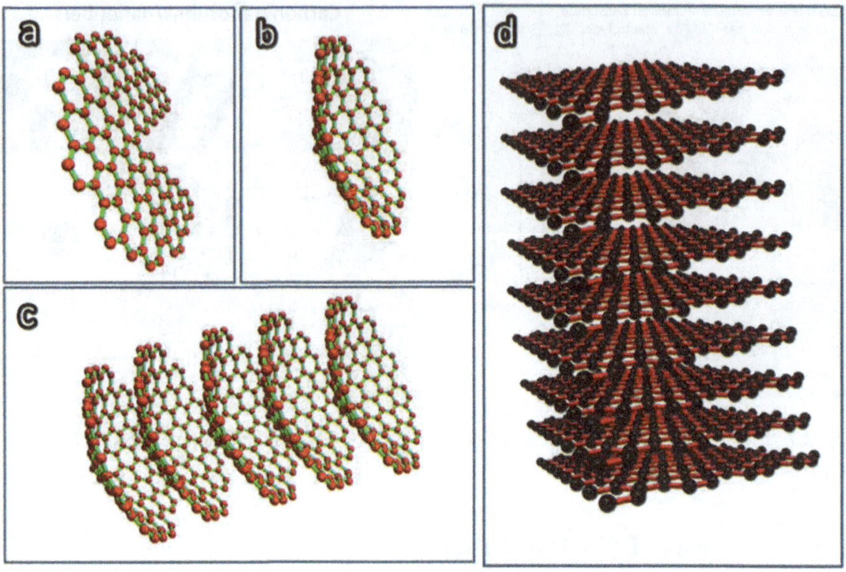

**Fig. 2.8** **a–c** Development of cup-stacked and CF platelet structure. Reproduced with permission from [80], copyright 2014, Materials

polymer solution and the specific processing settings. PAN and PET are the most often utilized polymers. Moreover, the application of polybenzimidazole (PBI), polyimides (PIs), poly(vinyl alcohol) (PVA), phenolic resin, poly(vinylidene fluoride) (PVDF), and lignin was demonstrated [81]. Following the successful production of the polymer nanofibers, a process involving heat will be employed to carbonize them, leading to the creation of CFs. The heat treating process parameters, including temperature, crystallinity, diameters, environment, and control, directly influence the morphology, purity, and porosity of the material. After the successful production of polymer nanofibers, the carbonization process will entail exposing the polymer nanofiber to a temperature of 1000 °C in a carefully regulated environment. Typically, the carbonization process leads to changes in volume and weight, resulting in a reduction in the diameter of the CNFs. Typically, CFs produced by the electrospinning technique have a tendency to form interconnected networks or sheet-like structures. This structure is suitable for battery and supercapacitor electrodes. The efficiency of batteries is mostly determined by the transit of ions in the electrolyte. Therefore, optimizing the pore structure is crucial for enhancing the efficiency of batteries [81].

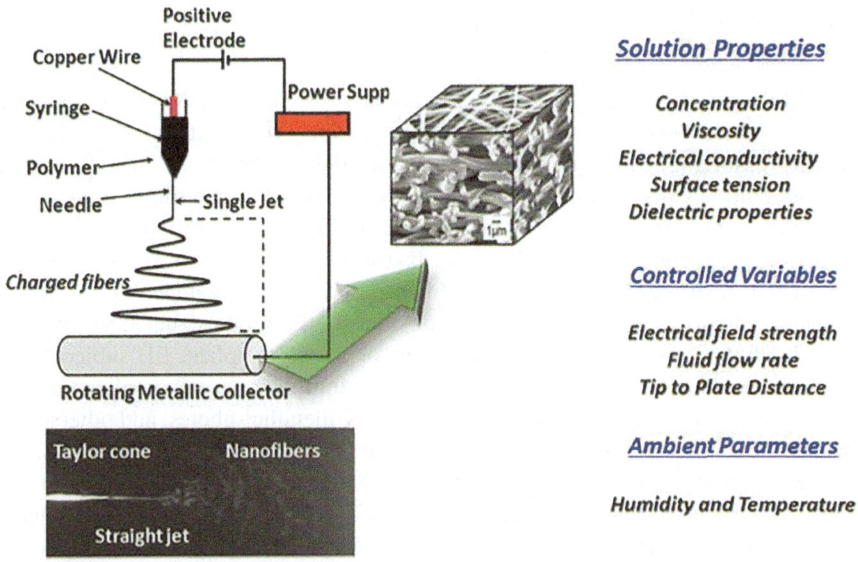


**Fig. 2.9** Electrospinning method for synthesis of CF. Reproduced with permission from [83]

## 2.4 Carbon Spheres

Typically, the methods for creating CSs may be divided into two categories: hard templating and soft templating. In the hard-templating technique, stiff particles that have been carefully prepared are used as "hard" core templates. These particles are sacrificed after a C-shell is formed around them, resulting in the creation of a hollow structure. The soft-templating process entails the self-assembly of C precursors and other organic molecules to build a hollow structure, with the purpose of arranging the structure as a whole. In this context, the primary structures often consist of small particles of "soft" precursor molecules, organic additives, or surfactants which will be eliminated during the ultimate pyrolysis process, resulting in the production of CSs [84–90]. The primary benefit of hard templates is the enhanced product control they provide. However, the removal of these templates is sometimes challenging and necessitates a technique that is not environmentally friendly. On the other hand, removing the soft template is rather simple, but it is more challenging to regulate the shape of the final goods. Various synthetic processes, such as hydrothermal carbonization (HTC), CVD, arc discharge, and laser ablation, can be used depending on the specific templates and C precursors involved [85, 91–93].

### 2.4.1 Hard-Templating Strategy

In hard-templating pathways, stiff particles are typically employed as the core template. These can be removed, based on their chemical composition, using methods such as dissolving, calcination, or etching after the formation of C shells around the core components. The CSs generated preserve the form and dimensions of the templates, functioning as inverse reproductions of them. Therefore, the characteristics of the resultant spheres, such as the distribution of diameters and the average size of the inner diameter, may be modified based on the parameters of the rigid templates employed [94, 95]. The typical method for synthesizing CSs involves the following steps: (i) applying a C precursor to the template, (ii) subjecting it to pyrolysis, and (iii) eliminating the template. Multiple types of templates can be utilized, including polymer spheres, silica spheres, metallic spheres, and others. The synthesis parameters may be modified to continuously alter the diameter of silica spheres, spanning from nanometer-scale (10–20 nm) to millimeter-scale (2–8 mm) spherical particles. The hollow carbon spheres (HCSs) generated from silica spheres retain the dimensions and morphology of the original templates. Because of the elevated temperatures necessary for the carbonization process, these carbonaceous structures exhibit a mostly hydrophobic and aromatic surface nature, which might hinder the extensive utilization of HCSs in certain catalytic applications. Titirici et al. successfully produced functionalized HCSs using hydrothermal carbonization (HTC) at low temperatures by modifying the polarity of the silica template surface, see Fig. 2.10i–ii [84]. Researchers found that using hydrophobic silica templates led to the creation of mesoporous C monolithic frameworks. On the other hand, when a silica template with intermediate hydrophobicity was used, C hollow spheres with a granular surface were produced. The addition of functional groups greatly improved the hydrophilic characteristics of the HCSs and elevated their stability in aqueous conditions.

### 2.4.2 Soft-Templating Strategy

The soft-templating approaches are appealing due to the ease of removing the related templates. The core templates might consist of colloidal systems, including micelles, emulsion droplets, gas bubbles, or vesicles. These systems often arise from precursor molecules or additives that were first introduced to construct the hollow structures. Despite the benefits, the morphology and uniformity of the hollow products obtained in their current state are not as desirable as those achieved by hard templating. This is mostly due to the variation in size and the dynamic nature of the soft templates [84, 93, 96, 97].

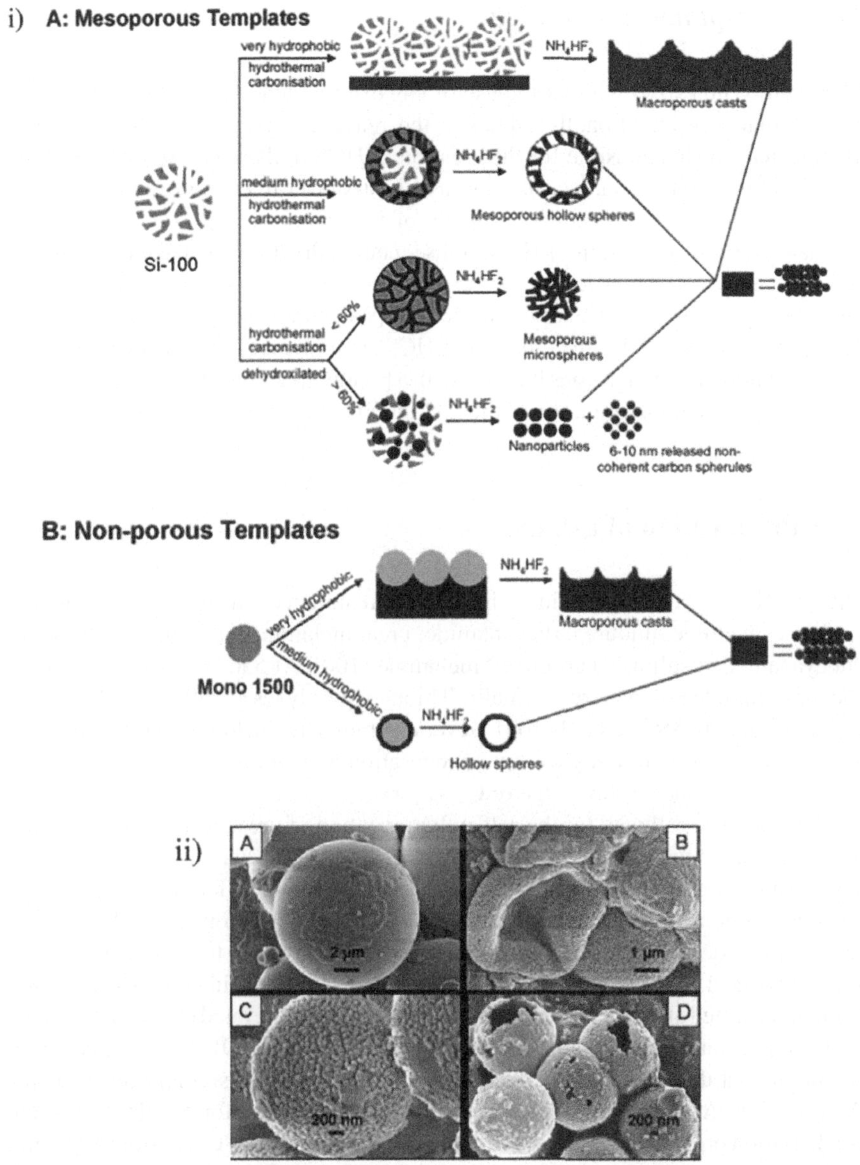


**Fig. 2.10** **i** Hydrothermal carbonization method in the existence of silica templates. Reproduced with permission from [88], Copyright 2016, Journal of material chemistry A. **ii** SEM images after experiments were completed with moderately hydrophobic silica templates. Reproduced with permission from [84], Copyright 2007, John Wiley

### 2.4.3 *Template-Free Strategy*

HCS were synthesized without the use of any external template, whether it be hard or soft, that was intentionally added to the reaction mixture and then removed. The mechanisms responsible for the creation of HCS in these conditions have been proposed but remain a subject of ongoing discussion. Liu et al. hypothesized that, in their experimental procedure, droplets of iron amine complexes first acted as templates for the generation of HCS. Subsequently, the droplets leaked out through faults and gaps in the shell, leading to the production of hollow spheres. This approach might be regarded as a self-templating strategy [98]. Nevertheless, Niwase et al. have out a potential process for the creation of HCSs that involves two distinct stages: (i) the generation of carbonaceous bubbles with a hollow spherical structure in the liquid phase and (ii) the transformation of these C-based bubbles into a graphitic form [99].

## 2.5 Preparation of g-$C_3N_4$

The g-$C_3N_4$ may be easily produced by thermal treatment of nitrogen-rich precursors, such as cyanamide, thiourea, dicyanamide, urea, melamine, and others ammonium thiocyanate, and sulfuric acid mixed melamine [100]. At 550 °C, cyanamide, urea, and dicyanamide polymerize thermally. Thiourea pyrolyzes at 450–650 °C, whereas melamine at 500–580 °C [101–103]. Several parameters, including the modification method, preparation flaws, packing, condensation temperature, and precursor type, can influence the adsorption edge of $C_3N_4$ [100].

The synthesis pathway for the self-polymerization of urea closely resembles that of the precursor for thiourea. Analogously, a shared characteristic is the release of ammonia during the process of polymerization. The main difference resides in the fact that thiourea condensation involves the extraction of hydrogen sulfide (HS) and carbon disulfide ($CS_2$), whereas urea polymerization involves the expulsion of $CO_2$ and $H_2O$ (Fig. 2.11). The gases generated during this heat-treating procedure, such as ammonia emitted at temperatures lower 200 °C or $CO_2$ at elevated temperatures, have a vital impact on the efficient processing of porous g-$C_3N_4$ [100]. Another distinction lies in the fact that their polymerization processes require dissimilar intermediates. Wang et al. were the first to employ cyanamide as a precursor for g-$C_3N_4$ [104]. The condensation process occurred at temperatures of 234 and 203 °C, yielding melamine and dicyandiamide, respectively. Subsequently, the condensation process resulted in the elimination of $NH_3$ gas. At a temperature of 335 °C, all the derivatives derived from melamine were detected. The synthesis of tri-s-triazine units was achieved by increasing the temperature to 390 °C, resulting in the rearrangement of melamine. The application of sustained heat and condensation at a temperature of 520 °C resulted in the creation of a polymeric variant of g-$C_3N_4$. However, the polymer became unstable around 600 °C. The polymer structure of g-$C_3N_4$ completely decomposed

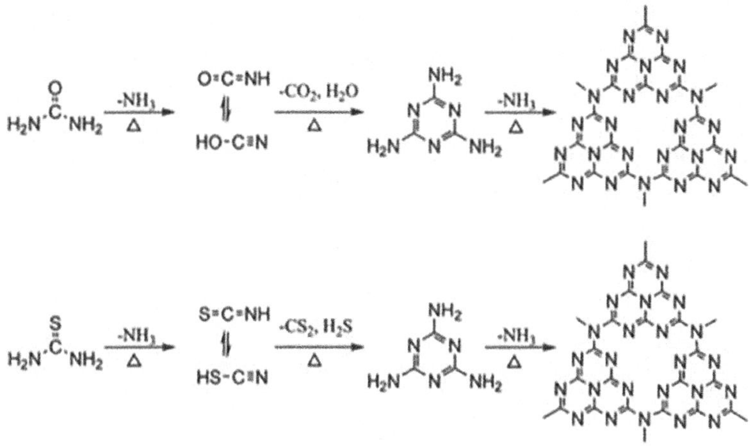


**Fig. 2.11** Self-polymerization of urea and thiourea in air to prepare g-$C_3N_4$. Reproduced with permission from [105], Copyright 2012, Journal of materials chemistry A

at temperatures exceeding 700 °C, forming cyano and nitrogen fragments with no residual material.

The quest for further innovative precursors resulted in the findings of guanidinium-based precursors. Schnick et al. synthesized g-$C_3N_4$ by heat condensing tricyanomelaminate and guanidinium dicyanamide [106]. It has been verified that melamine is a significant compound formed throughout the production process. Therefore, compounds derived from guanidinium can be transformed into the fundamental component of g-$C_3N_4$ and subsequently exposed to heat treating to undergo cross-linking. Figure 2.12 shows Long et al. using heat-induced desulfurization and condensation to convert guanidinium thiocyanate into thiocyanic acid and guanidine [107]. Both products reacted simultaneously, producing numerous outcomes. Guanidinium thiocyanate is cheap and available. Shi et al. employed an alternative chemical derived from guanidine to synthesize mesoporous g-$C_3N_4$ [108]. The synthesis used guanidine hydrochloride at 650 °C. The polymerization process was triggered by the application of heat. The chemical route depicted in.

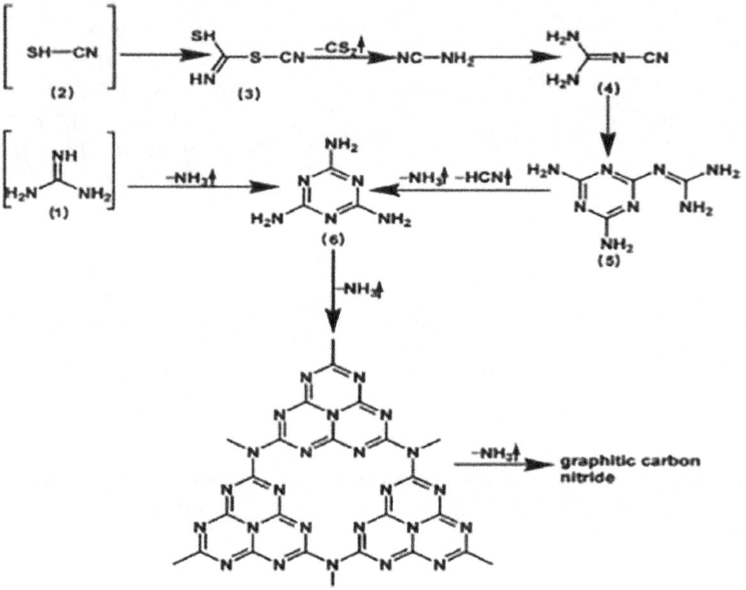


**Fig. 2.12** Self-polymerization of guanidine thiocyanate to the g-$C_3N_4$ network at high temperature [107] Copyright 2014, Journal of material chemistry A

## References

1. Zhang, Y., Small, J.P., Pontius, W.V., Kim, P.: Fabrication and electric-field-dependent transport measurements of mesoscopic graphite devices. Appl. Phys. Lett. **86**, 1–3 (2005). https://doi.org/10.1063/1.1862334
2. Sun, B., Pang, J., Cheng, Q., et al.: Synthesis of wafer-scale graphene with chemical vapor deposition for electronic device applications. Adv. Mater. Technol. **6**, (2021)
3. Rao, C.N.R., Maitra, U., Matte, H.S.S.R.: Synthesis, characterization, and selected properties of graphene. In: Graphene: Synthesis, Properties, and Phenomena, pp. 1–47. American Chemical Society (2012)
4. Bernhardt, T.M., Kaiser, B., Rademann, K.: Formation of superperiodic patterns on highly oriented pyrolytic graphite by manipulation of nanosized graphite sheets with the STM tip. Surf. Sci. **408**, 86–94 (1998). https://doi.org/10.1016/S0039-6028(98)00152-6
5. Lu, X., Yu, M., Huang, H., Ruoff, R.S.: Tailoring graphite with the goal of achieving single sheets. Nanotechnology **10**, 269–272 (1999). https://doi.org/10.1088/0957-4484/10/3/308
6. Ebbesen, T.W., Hiura, H.: Graphene in 3-dimensions: towards graphite origami. Adv. Mater. **7**, 582–586 (1995). https://doi.org/10.1002/adma.19950070618
7. Hiura, H., Ebbesen, T.W., Fujita, J., et al.: Role of $sp^3$ defect structures in graphite and carbon nanotubes. Nature **367**, 148–151 (1994). https://doi.org/10.1038/367148a0
8. Bhuyan, M.S.A., Uddin, M.N., Islam, M.M., et al.: Synthesis of graphene. Int. Nano Lett. **6**, 65–83 (2016)
9. Song, L., Ci, L., Gao, W., Ajayan, P.M.: Transfer printing of graphene using gold film. ACS Nano **3**, 1353–1356 (2009). https://doi.org/10.1021/nn9003082
10. Casiraghi, C., Hartschuh, A., Lidorikis, E., et al.: Rayleigh imaging of graphene and graphene layers. Nano Lett. **7**, 2711–2717 (2007). https://doi.org/10.1021/nl071168m
11. Su, C.Y., Lu, A.Y., Xu, Y., et al.: High-quality thin graphene films from fast electrochemical exfoliation. ACS Nano **5**, 2332–2339 (2011). https://doi.org/10.1021/nn200025p

12. Liu, J., Poh, C.K., Zhan, D., et al.: Improved synthesis of graphene flakes from the multiple electrochemical exfoliation of graphite rod. Nano Energy **2**, 377–386 (2013). https://doi.org/10.1016/j.nanoen.2012.11.003
13. Fan, X., Peng, W., Li, Y., et al.: Deoxygenation of exfoliated graphite oxide under alkaline conditions: a green route to graphene preparation. Adv. Mater. **20**, 4490–4493 (2008). https://doi.org/10.1002/adma.200801306
14. Abdelkader, A.M., Cooper, A.J., Dryfe, R.A.W., Kinloch, I.A.: How to get between the sheets: a review of recent works on the electrochemical exfoliation of graphene materials from bulk graphite. Nanoscale **7**, 6944–6956 (2015)
15. Parvez, K., Wu, Z.S., Li, R., et al.: Exfoliation of graphite into graphene in aqueous solutions of inorganic salts. J. Am. Chem. Soc. **136**, 6083–6091 (2014). https://doi.org/10.1021/ja5017156
16. Wan, H., Wei, C., Zhu, K., et al.: Preparation of graphene sheets by electrochemical exfoliation of graphite in confined space and their application in transparent conductive films. ACS Appl. Mater. Interfaces **9**, 34456–34466 (2017). https://doi.org/10.1021/acsami.7b09891
17. Achee, T.C., Sun, W., Hope, J.T., et al.: High-yield scalable graphene nanosheet production from compressed graphite using electrochemical exfoliation. Sci. Rep. **8**, (2018). https://doi.org/10.1038/s41598-018-32741-3
18. Lee, X.J., Hiew, B.Y.Z., Lai, K.C., et al.: Review on graphene and its derivatives: synthesis methods and potential industrial implementation. J. Taiwan Inst. Chem. Eng. **98**, 163–180 (2019). https://doi.org/10.1016/j.jtice.2018.10.028
19. Mao, S., Pu, H., Chen, J.: Graphene oxide and its reduction: modeling and experimental progress. RSC Adv. **2**, 2643–2662 (2012)
20. Muzyka, R., Kwoka, M., Smędowski, Ł, et al.: Oxidation of graphite by different modified Hummers methods. Xinxing Tan Cailiao/New Carbon Mater. **32**, 15–20 (2017). https://doi.org/10.1016/S1872-5805(17)60102-1
21. Das, B., Kundu, R., Chakravarty, S.: Preparation and characterization of graphene oxide from coal. Mater. Chem. Phys. **290**, (2022). https://doi.org/10.1016/j.matchemphys.2022.126597
22. Brodie, B.C.: Sur le poids atomique du graphite. Google Scholar
23. Staudenmaier, L.: Verfahren zur Darstellung der Graphitsäure. Berichte der Dtsch Chem Gesellschaft **31**, 1481–1487 (1898). https://doi.org/10.1002/cber.18980310237
24. Brodie, B.C.: Sur le poids atomique. Google Scholar
25. Shin, H.J., Kim, K.K., Benayad, A., et al.: Efficient reduction of graphite oxide by sodium borohydride and its effect on electrical conductance. Adv. Funct. Mater. **19**, 1987–1992 (2009). https://doi.org/10.1002/adfm.200900167
26. Amarnath, C.A., Hong, C.E., Kim, N.H., et al.: Efficient synthesis of graphene sheets using pyrrole as a reducing agent. Carbon N. Y. **49**, 3497–3502 (2011). https://doi.org/10.1016/j.carbon.2011.04.048
27. Pham, V.H., Cuong, T.V., Nguyen-Phan, T.D., et al.: One-step synthesis of superior dispersion of chemically converted graphene in organic solvents. Chem. Commun. **46**, 4375–4377 (2010). https://doi.org/10.1039/c0cc00363h
28. Zhang, J., Yang, H., Shen, G., et al.: Reduction of graphene oxide vial-ascorbic acid. Chem. Commun. **46**, 1112–1114 (2010). https://doi.org/10.1039/b917705a
29. Zhou, X., Zheng, J., Wu, H., et al.: Reducing graphene oxide via hydroxylamine: a simple and efficient route to graphene. J. Phys. Chem. C **115**, 11957–11961 (2011). https://doi.org/10.1021/jp202575j
30. Zhu, C., Guo, S., Fang, Y., et al.: Reducing sugar: new functional molecules for the green synthesis of graphene nanosheets. ACS Nano **4**, 2429–2437 (2010). https://doi.org/10.1021/nn1002387
31. Kumar, N., Salehiyan, R., Chauke, V., et al.: Top-down synthesis of graphene: a comprehensive review. Flat Chem. **27**, 100224 (2021). https://doi.org/10.1016/J.FLATC.2021.100224
32. Hintsho, N., Shaikjee, A., Masenda, H., et al.: Direct synthesis of carbon nanofibers from South African coal fly ash. Nanoscale Res. Lett. **9**, 1–11 (2014). https://doi.org/10.1186/1556-276X-9-387

33. Hlekelele, L., Franklyn, P.J., Dziike, F., Durbach, S.H.: $TiO_2$ composited with carbon nanofibers or nitrogen-doped carbon nanotubes synthesized using coal fly ash as a catalyst: bisphenol-A photodegradation efficiency evaluation. New J. Chem. **42**, 4531–4542 (2018). https://doi.org/10.1039/c7nj04787h
34. Lim, J.Y., Mubarak, N.M., Abdullah, E.C., et al.: Recent trends in the synthesis of graphene and graphene oxide based nanomaterials for removal of heavy metals—a review. J. Ind. Eng. Chem. **66**, 29–44 (2018)
35. Muñoz, R., Gómez-Aleixandre, C.: Review of CVD synthesis of graphene. Chem. Vap. Depos. **19**, 297–322 (2013)
36. Somani, P.R., Somani, S.P., Umeno, M.: Planer nano-graphenes from camphor by CVD. Chem. Phys. Lett. **430**, 56–59 (2006). https://doi.org/10.1016/j.cplett.2006.06.081
37. Lee, J.H., Lee, E.K., Joo, W.J., et al.: Wafer-scale growth of single-crystal monolayer graphene on reusable hydrogen-terminated germanium. Science **344**, 286–289 (2014). https://doi.org/10.1126/science.1252268
38. Chandrashekar, B.N., Deng, B., Smitha, A.S., et al.: Roll-to-roll green transfer of CVD graphene onto plastic for a transparent and flexible triboelectric nanogenerator. Adv. Mater. **27**, 5210–5216 (2015). https://doi.org/10.1002/adma.201502560
39. Kobayashi, T., Bando, M., Kimura, N., et al.: Production of a 100-m-long high-quality graphene transparent conductive film by roll-to-roll chemical vapor deposition and transfer process. Appl. Phys. Lett. **102**, (2013). https://doi.org/10.1063/1.4776707
40. Polsen, E.S., McNerny, D.Q., Viswanath, B., et al.: High-speed roll-to-roll manufacturing of graphene using a concentric tube CVD reactor. Sci. Rep. **5**, 1–12 (2015). https://doi.org/10.1038/srep10257
41. Chenxing, W.: Growth of epitaxial graphene on SiC (0001) by thermal decomposition in argon and its etching in oxygen atmosphere. nagoya.repo.nii.ac.jp (2016)
42. Forbeaux, I., Themlin, J., Debever, J.: Heteroepitaxial graphite on interface formation through conduction-band electronic structure. Phys. Rev. B—Condens. Matter Mater. Phys. **58**, 16396–16406 (1998). https://doi.org/10.1103/PhysRevB.58.16396
43. Ruan, M., Hu, Y., Guo, Z., et al.: Epitaxial graphene on silicon carbide: introduction to structured graphene. MRS Bull. **37**, 1138–1147 (2012). https://doi.org/10.1557/mrs.2012.231
44. Berger, C., Song, Z., Li, T., et al.: Ultrathin epitaxial graphite: 2D electron gas properties and a route toward graphene-based nanoelectronics. J. Phys. Chem. B **108**, 19912–19916 (2004). https://doi.org/10.1021/jp040650f
45. Saito, T., Matsushige, K., Tanaka, K.: Chemical treatment and modification of multi-walled carbon nanotubes. Phys. B: Condens. Matter. 280–283 (2002)
46. Kosynkin, D.V., Higginbotham, A.L., Sinitskii, A., et al.: Longitudinal unzipping of carbon nanotubes to form graphene nanoribbons. Nature **458**, 872–876 (2009). https://doi.org/10.1038/nature07872
47. Bethune, D.S., Klang, C.H., De Vries, M.S., et al.: Cobalt-catalysed growth of carbon nanotubes with single-atomic-layer walls. Nature **363**, 605–607 (1993). https://doi.org/10.1038/363605a0
48. Moosa, A.A., Abed, M.S.: Graphene preparation and graphite exfoliation. Turkish J. Chem. **45**, 493–519 (2021)
49. Iijima, S.: Growth of carbon nanotubes. Mater. Sci. Eng. B **19**, 172–180 (1993). https://doi.org/10.1016/0921-5107(93)90184-O
50. Iijima, S., Ichihashi, T.: Single-shell carbon nanotubes of 1-nm diameter. Nature **363**, 603–605 (1993). https://doi.org/10.1038/363603a0
51. Arora, N., Sharma, N.N.: Arc discharge synthesis of carbon nanotubes: comprehensive review. Diam. Relat. Mater. **50**, 135–150 (2014)
52. Amelinckx, S., Bernaerts, D., Zhang, X.B., et al.: A structure model and growth mechanism for multishell carbon nanotubes. Science **267**, 1334–1338 (1995). https://doi.org/10.1126/science.267.5202.1334
53. Karthikeyan, S., Mahalingam, P., Karthik, M.: Large scale synthesis of carbon nanotubes. E-J. Chem. **6**, 1–12 (2009)

54. Gamaly, E.G., Ebbesen, T.W.: Mechanism of carbon nanotube formation in the arc discharge. Phys. Rev. B **52**, 2083–2089 (1995). https://doi.org/10.1103/PhysRevB.52.2083
55. Wang, Y.H., Chiu, S.C., Lin, K.M., Li, Y.Y.: Formation of carbon nanotubes from polyvinyl alcohol using arc-discharge method. Carbon N. Y. **42**, 2535–2541 (2004). https://doi.org/10.1016/j.carbon.2004.05.028
56. Zhang, M., Kelleher, E.J.R., Runcorn, T.H., et al.: Mid-infrared Raman-soliton continuum pumped by a nanotube-mode-locked sub-picosecond Tm-doped MOPFA. Opt. Express **21**, 23261 (2013). https://doi.org/10.1364/oe.21.023261
57. Thompson, D.: Nanotechnology: basic science and emerging technologies. Gold Bull. **35**, 135–136 (2002). https://doi.org/10.1007/bf03214856
58. Ashraf, A., Yaqub, K., Javeed, S., et al.: Sublimation of graphite in continuous and pulsed arc discharges. Turk. J. Phys. **34**, 33–42 (2010). https://doi.org/10.3906/fiz-0903-10
59. Chernysheva, M., Rozhin, A., Fedotov, Y., et al.: Carbon nanotubes for ultrafast fibre lasers. Nanophotonics **6**, 1–30 (2017)
60. Ebbesen, T.W., Nature, P.M.: Large-scale synthesis of carbon nanotubes. Nature (1992)
61. Sinnott, S.B., Andrews, R.: Carbon nanotubes: synthesis, properties, and applications. Crit. Rev. Solid State Mater. Sci. **26** (2001)
62. Varshney, K.: Carbon nanotubes: a review on synthesis, properties and applications. Int. J. Eng. Res. Gen. Sci. **2**
63. Hernadi, K., Fonseca, A., Nagy, J.B., Bernaerts, D.: Catalytic Synthesis of Carbon Nanotubes, pp. 81–97. www-g.eng.cam.ac.uk (1998)
64. José-Yacamán, M., Miki-Yoshida, M., Rendón, L., Santiesteban, J.G.: Catalytic growth of carbon microtubules with fullerene structure. Appl. Phys. Lett. **62**, 657–659 (1993). https://doi.org/10.1063/1.108857
65. Pezzin, S.H., Amico, S.C., Coelho, L.A.F., de Andrade, M.J.: Nanoreinforcements for nanocomposite materials. In: Nanostructured Materials for Engineering Applications, pp. 119–131. Springer, Berlin Heidelberg (2011)
66. Seo, J.W., Magrez, A., Milas, M., et al.: Catalytically grown carbon nanotubes: from synthesis to toxicity. J. Phys. D: Appl. Phys. **40**, (2007). https://doi.org/10.1088/0022-3727/40/6/R01
67. Teo, K.B.K., Singh, C., et al.: Catalytic synthesis of carbon nanotubes and nanofibers. www-g.eng.cam.ac.uk (2003)
68. Paradise, M., Goswami, T.: Carbon nanotubes–production and industrial applications. Mater. Des. (2007)
69. Journet, C., Bernier, P.: Production of carbon nanotubes. Appl. Phys. A Mater. Sci. Process. **67**, 1–9 (1998). https://doi.org/10.1007/s003390050731
70. Szabó, A., Perri, C., Csató, A., et al.: Synthesis methods of carbon nanotubes and related materials. Materials (Basel) **3**, 3092–3140 (2010). https://doi.org/10.3390/ma3053092
71. Hsu, W.K., Hare, J.P., Terrones, M., et al.: Condensed-phase nanotubes. nature.com
72. Bai, J.B., Hamon, A.L., Marraud, A., et al.: Synthesis of SWNTs and MWNTs by a molten salt (NcCl) method. Chem. Phys. Lett. **365**, 184–188 (2002). https://doi.org/10.1016/S0009-2614(02)01447-1
73. Harris, P.J.F.: Solid state growth mechanisms for carbon nanotubes. Carbon N. Y. **45**, 229–239 (2007)
74. Novoselova, I.A., Oliinyk, N.F., Volkov, S.V., et al.: Electrolytic synthesis of carbon nanotubes from carbon dioxide in molten salts and their characterization. Phys. E: Low-Dimens. Syst. Nanostructures **40**, 2231–2237 (2008). https://doi.org/10.1016/j.physe.2007.10.069
75. Grobert, N.: Carbon nanotubes–becoming clean. Mater. Today (2007)
76. Liu, X., Licht, G., Wang, X., Licht, S.: Controlled transition metal nucleated growth of carbon nanotubes by molten electrolysis of $CO_2$. Catalysts **12**, 137 (2022). https://doi.org/10.3390/catal12020137
77. Ge, M., Sattler, K.: Observation of fullerene cones. Chem. Phys. Lett. **220**, 192–196 (1994). https://doi.org/10.1016/0009-2614(94)00167-7
78. De Jong, K.P., Geus, J.W.: Carbon nanofibers: catalytic synthesis and applications. Catal. Rev. Sci. Eng. **42**, 481–510 (2000). https://doi.org/10.1081/CR-100101954

79. Kim, Y.A., Hayashi, T., Endo, M., Dresselhaus, M.S.: Google Scholar
80. Feng, L., Xie, N., Zhong, J.: Carbon nanofibers and their composites: a review of synthesizing, properties and applications. Materials (Basel) **7**, 3919–3945 (2014)
81. Inagaki, M., Yang, Y., Kang, F.: Carbon nanofibers prepared via electrospinning. Adv. Mater. **24**, 2547–2566 (2012)
82. Zhang, L., Aboagye, A., Kelkar, A., et al.: A review: carbon nanofibers from electrospun polyacrylonitrile and their applications. J. Mater. Sci. **49**, 463–480 (2014)
83. Gopiraman, M., Soo Kim, I.: Preparation, characterization, and applications of electrospun carbon nanofibers and its composites. In: Electrospinning and Electrospraying—Techniques and Applications. IntechOpen (2019)
84. Titirici, M.M., Thomas, A., Antonietti, M.: Replication and coating of silica templates by hydrothermal carbonization. Adv. Funct. Mater. **17**, 1010–1018 (2007). https://doi.org/10.1002/adfm.200600501
85. Jin, Y.Z., Gao, C., Hsu, W.K., et al.: Large-scale synthesis and characterization of carbon spheres prepared by direct pyrolysis of hydrocarbons. Carbon N. Y. **43**, 1944–1953 (2005). https://doi.org/10.1016/j.carbon.2005.03.002
86. Choma, J., Jamioła, D., Augustynek, K., et al.: New opportunities in Stöber synthesis: preparation of microporous and mesoporous carbon spheres. J. Mater. Chem. **22**, 12636–12642 (2012). https://doi.org/10.1039/c2jm31678a
87. Panickar, R., Sobhan, C.B., Chakravorti, S.: Chemical vapor deposition synthesis of carbon spheres: effects of temperature and hydrogen. Vacuum **172**, (2020). https://doi.org/10.1016/j.vacuum.2019.109108
88. Li, S., Pasc, A., Fierro, V., Celzard, A.: Hollow carbon spheres, synthesis and applications—a review. J. Mater. Chem. A **4**, 12686–12713 (2016)
89. Wang, X., Feng, J., Bai, Y., et al.: Synthesis, properties, and applications of hollow micro-/nanostructures. Chem. Rev. **116**, 10983–11060 (2016)
90. Li, Y., Shi, J.: Hollow-structured mesoporous materials: chemical synthesis, functionalization and applications. Adv. Mater. **26**, 3176–3205 (2014)
91. Mezzavilla, S., Baldizzone, C., Mayrhofer, K.J.J., Schüth, F.: General method for the synthesis of hollow mesoporous carbon spheres with tunable textural properties. ACS Appl. Mater. Interfaces **7**, 12914–12922 (2015). https://doi.org/10.1021/acsami.5b02580
92. Wang, J., Chen, Q., Liu, X., et al.: Hard-templating synthesis of mesoporous carbon spheres with controlled particle size and mesoporous structure for enzyme immobilization. Mater. Chem. Phys. **129**, 1035–1041 (2011). https://doi.org/10.1016/j.matchemphys.2011.05.085
93. Liu, J., Yang, T., Wang, D.W., et al.: A facile soft-template synthesis of mesoporous polymeric and carbonaceous nanospheres. Nat. Commun. **4**, 1–7 (2013). https://doi.org/10.1038/ncomms3798
94. Kubo, S., Demir-Cakan, R., Zhao, L., et al.: Porous carbohydrate-based materials via hard templating. Chem. Sus. Chem. **3**, 188–194 (2010). https://doi.org/10.1002/cssc.200900126
95. Xia, Y., Gates, B., Yin, Y., Lu, Y.: Monodispersed colloidal spheres: old materials with new applications. Adv. Mater. **12**, 693–713 (2000). https://doi.org/10.1002/(SICI)1521-4095(200005)12:10%3c693::AID-ADMA693%3e3.0.CO;2-J
96. Javed, H., Pani, S., Antony, J., et al.: Synthesis of mesoporous carbon spheres: via a soft-template route for catalyst supports in PEMFC cathodes. Soft Matter **17**, 7743–7754 (2021). https://doi.org/10.1039/d1sm00450f
97. Tian, H., Liang, J., Liu, J.: Nanoengineering carbon spheres as nanoreactors for sustainable energy applications. Adv. Mater. **31**, (2019). https://doi.org/10.1002/adma.201903886
98. Liu, B., Jia, D., Meng, Q., Rao, J.: A novel method for preparation of hollow carbon spheres under a gas pressure atmosphere. Carbon N. Y. **45**, 668–670 (2007)
99. Li, G., Xu, L., Hao, Q., et al.: Synthesis, characterization and application of carbon nanocages as anode materials for high-performance lithium-ion batteries. RSC Adv. **2**, 284–291 (2012). https://doi.org/10.1039/c1ra00631b
100. Ong, W.J., Tan, L.L., Ng, Y.H., et al.: Graphitic carbon nitride (g-$C_3N_4$)-based photocatalysts for artificial photosynthesis and environmental remediation: are we a step closer to achieving sustainability? Chem. Rev. **116**, 7159–7329 (2016)

101. Cao, S., Low, J., Yu, J., Jaroniec, M.: Polymeric photocatalysts based on graphitic carbon nitride. Adv. Mater. **27**, 2150–2176 (2015). https://doi.org/10.1002/adma.201500033
102. Hao, Q., Jia, G., Wei, W., et al.: Graphitic carbon nitride with different dimensionalities for energy and environmental applications. Nano Res. **13**, 18–37 (2020)
103. Wang, S., Li, J., Li, Q., et al.: Metal single-atom coordinated graphitic carbon nitride as an efficient catalyst for CO oxidation. Nanoscale **12**, 364–371 (2020). https://doi.org/10.1039/c9nr07726j
104. Wang, X., Maeda, K., Thomas, A., et al.: A metal-free polymeric photocatalyst for hydrogen production from water under visible light. Nat. Mater. **8**, 76–80 (2009). https://doi.org/10.1038/nmat2317
105. Zhang, G., Zhang, J., Zhang, M., Wang, X.: Polycondensation of thiourea into carbon nitride semiconductors as visible light photocatalysts. J. Mater. Chem. **22**, 8083–8091 (2012). https://doi.org/10.1039/c2jm00097k
106. Lotsch, B.V., Schnick, W.: Thermal conversion of guanylurea dicyanamide into graphitic carbon nitride via prototype CNx precursors. Chem. Mater. **17**, 3976–3982 (2005). https://doi.org/10.1021/cm050350q
107. Long, B., Lin, J., Wang, X.: Thermally-induced desulfurization and conversion of guanidine thiocyanate into graphitic carbon nitride catalysts for hydrogen photosynthesis. J. Mater. Chem. A **2**, 2942–2951 (2014). https://doi.org/10.1039/C3TA14339B
108. Shi, L., Liang, L., Wang, F., et al.: Polycondensation of guanidine hydrochloride into a graphitic carbon nitride semiconductor with a large surface area as a visible light photocatalyst. Catal. Sci. Technol. **4**, 3235–3243 (2014). https://doi.org/10.1039/c4cy00411f

# Chapter 3
# Dye Degradation Application

## 3.1 Water, A Basic Need

Water is essential for life on Earth, as it sustains living systems. It is vital to civilization evolution [1]. About 70.9% of Earth's surface is occupied by water, known as the hydrosphere, which houses most of the ecosystem's living creatures. Of the Earth's 70.9% area, 96.5% is salty water (e.g., oceans, estuaries, seas), whereas 1.7% is subterranean water, including polar glaciers and Greenland. Gleick et al. found that just 0.001% is present as vaporized clouds [2]. Freshwater is an essential resource crucial for the sustenance of living beings on our planet [3]. As per the United Nations assessment, the Earth contains enough water to form a layer that is 3000 m deep. Water quality, including chemical, physical, and biological factors, must be clean and safe for drinking and domestic use [4]. Water is essential for economic progress due to its qualities as a solvent and factory cooling. Safe drinking water awareness and treatment have improved globally in recent decades. According to world water research organizations, over 1 billion individuals now do not have access to potable water, and this scenario is projected to deteriorate by 2025 [5]. Water quality can be impacted by natural and human activity. Water composition is highly influenced by local geological circumstances. The World Health Organization (WHO) has established criteria for safe drinking water quality, including maximum acceptable biological and chemical contaminants (WHO 2004). WHO and UNICEF report that contaminated drinking water and poor hygiene cause 2.2 million deaths annually in developing countries [6], and 60% of newborns face life-threatening contagious and parasitic diseases [7]. Excessive use of inorganic fertilizers/pesticides in agriculture has increased soil, groundwater, and surface water pollution, harming human health [8, 9].



M. Ikram et al., *Carbon-Based Nanomaterials for Environmental Applications*,
Engineering Materials, https://doi.org/10.1007/978-3-031-59390-1_3

## 3.2 Water Pollution

Global aquatic reservoirs face pollution from various sources, including hydrocarbons, pesticides, drugs, cosmetics, chemicals, hormonal imbalances, and toxic heavy metals [10, 11]. Groundwater serves as a prevalent source of drinking water in both rural and urban regions, encompassing India where it caters to 95% of the rural population and 30–40% of the urban population. In recent decades, industrialization and unmanaged agriculture have introduced harmful contaminants to groundwater, including chemicals, metals, radionuclides, and organic reactants [12]. Inorganic pollutants often found in water include lead, fluoride, iron, nitrates, and heavy metals. When these contaminants exceed allowed levels, they compromise the drinking quality of water for living organisms. As a result of human actions, nitrates, phosphates, and heavy metals are added to waterways through things like poorly managed sewage systems, industry discharges, and poor farming practices. Arsenic, fluoride, and iron come from the ground. The primary determinants influencing the quality of groundwater are heavy industry, urbanization, societal modernization, and fast population growth [13].

Water contamination is mostly focused in industry, agriculture, natural causes, and inadequate supply and sewage treatment systems. Primarily, the largest source of water pollution is the industrial sector, including various businesses such as tanneries, pulp, distilleries, textile manufacturers, paper mills, iron and steel producers, food processing plants, nuclear facilities, and others. Industrial production can result in the emission of a wide range of harmful compounds, including both organic and inorganic molecules, as well as toxic solvents and volatile organic chemicals. Insufficiently treated discharge of these wastes into aquatic environments will result in water contamination [14]. Arsenic, cadmium, and chromium are essential contaminants released into wastewater, and the industrial sector plays a large role in the emission of these dangerous substances [15]. As urbanization has progressed, the amount of wastewater generated by industrial output has steadily risen [16]. Agriculture is intimately linked to water contamination. Agricultural pesticides, organic agricultural waste, and nitrogen fertilizers contribute to water pollution. Agricultural activities contaminate water sources with phosphate, nitrates, soil sediments, pesticides, salts, and pathogens [17]. Agriculture has had a profound negative impact on all pristine ecosystems of freshwater [18]. In water-scarce developing nations such as China and India, unfiltered or partly processed wastewater is used for agriculture. Sewage contaminants pose environmental and health problems. In developing nations such as China, the disparity in surface water resources has necessitated the prolonged utilization of wastewater irrigation to fulfill the water requirements for agricultural production. Consequently, this practice has led to significant contamination of agricultural land and food, including the presence of pesticide residues and heavy metals. These pollutants pose a threat to both food quality and the health of humans. Pesticides have detrimental effects on human health when ingested through the use of drinking water. Based on the Health Life Expectancy Longitudinal Survey, a 10% rise in pesticide usage resulted in a 1% increase in the medical disability index for those over the age of 65 [19]. The Musi River example in India demonstrates a greater prevalence of illness in areas that utilize wastewater for irrigation compared to those who use regular water. Furthermore, water contamination has a correlation with natural causes. Using the Child Loess Plateau as an illustration, the water quality in this area exhibits a greater concentration of trace elements compared to the global average. These trace elements originate from both natural weathering processes and industrial activities. The presence of elevated levels of salt and salinity in river water is linked to poor water quality, as indicated by a study conducted by Xiao et al. in 2019 [20]. Hexavalent chromium (Cr) contamination is the predominant kind of water contamination in the central region of the Loess Plateau. It arises from both natural factors and human activity. The primary contributors to surface water contamination are loess and mudstone, with an additional significant influence being the presence of groundwater with large levels of hexavalent Cr. Water supplies and sewage treatment facilities have a significant influence on the quality of drinking water, particularly in underdeveloped countries. China's fast economic expansion, urbanization, and industrialization along with underinvestment in basic water supply facilities have resulted in water pollution and increased exposure to parasites, infectious illnesses, industrial chemicals, heavy metals, and algae toxins [16, 21]. An economic model forecasts the impact of water purification technologies on water quality and public health. From 100 to 90% of home water treated using water filtration technology, health benefits drop by 96%. When pretreatment water quality is at danger, the drop is greater. Water contamination is caused by humans and nature. Water quality is affected by urbanization, population increase, industrial output, climatic change, and religious activities [22, 23]. Poor solid waste, sand, and gravel disposal can also degrade water quality [22].

## 3.3 Techniques for Water Treatment

To create novel wastewater treatment technologies and meet clean water criteria, progress is being made [24]. With current procedures, polluted discharged water has proven difficult to remediate [25]. The literature describes several wastewater treatment methods. Figure 3.1 shows that chemical, biological, and physical processes are helpful for water purification in numerous ways. Dye concentration, sewage composition, process cost, and wastewater contaminants affect their selection [26]. Each treatment procedure has advantages and disadvantages. Industrial applications seldom value treatment methods with large installation and operation costs, processing time, poor output, and harmful consequences [27]. Thus, an alternate treatment technique that totally degrades or removes pollutants is essential [28].

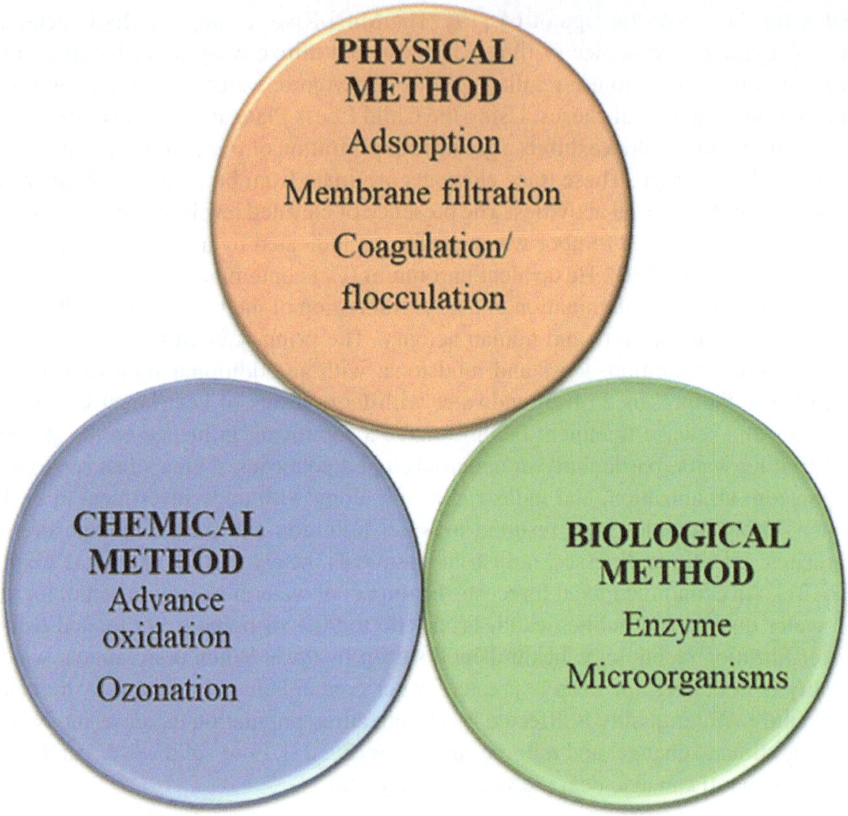


**Fig. 3.1** Conventional and advanced techniques used for wastewater treatment Reproduced with permission from reference [29], Copyright 202, springer

## 3.4 Physical Methods

The basic principle of physical pollution removal strategies is in the mass transfer approach [30]. Its simplicity, flexibility, efficiency, and pollutant recyclability make it increasingly popular [27, 31]. An additional benefit of this strategy is the reduced chemical demand. Physical therapy is considered more trustworthy than other therapies due to its independence from live organisms [30]. In recent years, researchers have increasingly utilized adsorption as a physical approach due to its notable efficacy and cost-effectiveness [32].

### 3.4.1 Adsorption

Adsorption is the binding of molecules or ions to a solid surface, either by chemical or physical means. The materials that serve as the surface are referred to as adsorbents, while the molecules or ions that are extracted from the phase of water are called adsorbates (Fig. 3.2). In contemporary times, the process of adsorption has shown to be advantageous for the purpose of water purification, since it possesses the capability to eliminate minute amounts of impurities from water [32, 33]. Adsorption is the most effective treatment method for wastewater containing dyes because it is inexpensive, has a straightforward design, and is resistant to dangerous pollutants. In addition, it should be noted that adsorption does not lead to the occurrence of secondary pollution, as demonstrated by Sanghi and Bhattacharya in 2002 and Carvalho et al. in 2011 [34, 35].

Adsorption methods are popular because they decolorize wastewater with several colors more efficiently [36]. Many dyes adsorb well on activated C. However, its

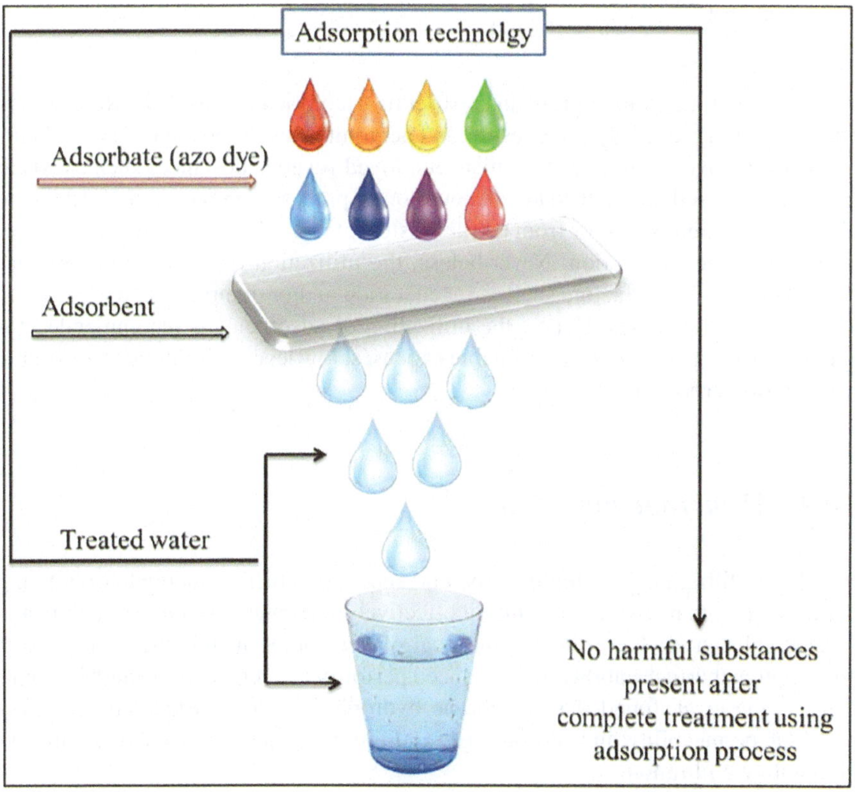


**Fig. 3.2** Diagram of adsorption process as a wastewater treatment. Reproduced with permission from [40], Copyright 2022, Springer

**Table 3.1** Various adsorbents for adsorption of dye

| Sr. no | Adsorbents | Dyes | References |
|---|---|---|---|
| 1 | Eichhornia charcoal and groundnut shell charcoal | Dye basic blue 9 | [41] |
| 2 | Waste tea-activated carbon | Acid blue 25 | [42] |
| 3 | Treated ginger waste | Crystal violet (dye | [43] |
| 4 | Straw-based absorbent | MB | [44] |
| 5 | Potato plant waste | MB and malachite green dye | [39] |
| 6 | Palm-trees waste and date stones | MB | [45] |
| 7 | *Capsicum annuum* seeds | Reactive blue 49 | [46] |
| 8 | Pistachio hull powder | MB | [47] |
| 9 | *Symphoricarpos albus*, modified with sodium diethyldithiocarbamate | Reactive dye | [48] |
| 10 | Modified wheat residue | Anionic dye (Reactive red-24) | [49] |
| 11 | Immobilized eggshell with a polymer mixture of alginate and polyvinyl alcohol | Remazol resactive red 198 | [50] |

high cost and regeneration problems restrict its decolorization use [37]. Researchers utilized peat, fly ash, polymeric resins, and bentonite clay as inexpensive adsorbents for cost-effective adsorption. Scientists employed several substances such as wheat residue, powdered nut shell charcoal, date stones, processed ginger water, and potato plant debris to remove color from textile waste water. Table 3.1 lists dyes and adsorbents for water purification. Nevertheless, the utilization of these adsorbents has been restricted due to factors such as regeneration and/or dumping, sludge creation, and exorbitant expenses. Hence, the utilization of adsorbents is recommended for applications characterized by affordable expenses, low levels of pollutants, or simple regeneration procedures [38, 39].

### *3.4.2 Membrane Filtration*

Membrane filtration is a highly efficient and cost-effective method for treating wastewater from many sectors, since it effectively separates and removes colors and organic substances. Nevertheless, the goal of developing membranes that possess both enough thermal stability and enhanced performance remains a formidable challenge. Yang et al. found that membrane hydrophilicity and surface charge affect rejection, permeability, and antifouling [51]. Figure 3.3 illustrates the overall process of membrane filtration.

Nanofiltration (NF) technique has been shown to be an effective technique for purifying and separating wastewater. Concentration polymerization and fouling can

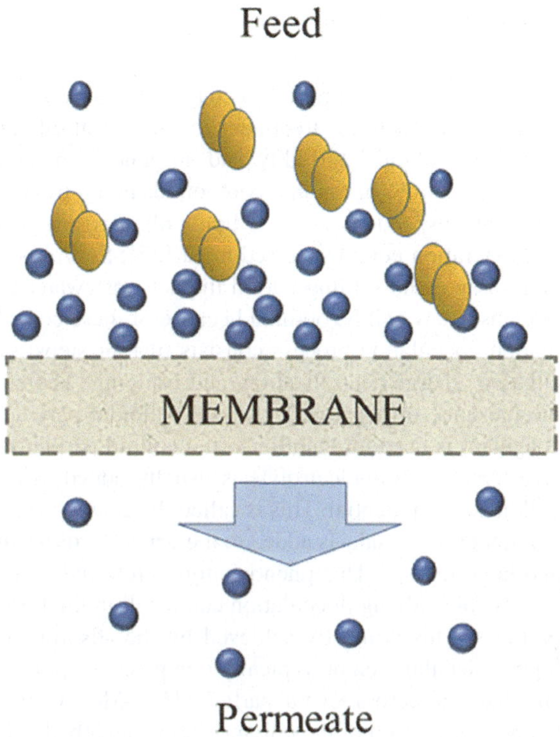


**Fig. 3.3** Membrane filtration mechanism Reproduced with permission from reference [29], Copyright 202, Springer

significantly limit NF membrane function. Membrane fouling reduces the permeability of the membrane by obstructing the pores with a coating of organic molecules. In addition, elevated salinity has a detrimental impact on membrane flow, since it increases osmotic pressure as a result of salt exclusion. Therefore, the use of a hydrophilic surface NF membrane is preferred for effectively treating large volumes of textile wastewater [52]. Typically, industrial-grade commercial NF membranes have tightly packed surface structures, resulting in a high ability to block salts and organic substances (such as dyes). However, this also leads to a decrease in the rate at which substances may pass through the membrane [53].

Ultrafiltration (UF) is employed in dyeing purposes and salt extraction processes due to its ability to enhance salt penetration and permeability flux. Comparative studies show that UF membranes have better separation efficiency because they allow salt to pass through more easily and can handle huge amounts of material due to their small osmotic pressure. Polyacrylonitrile (PAN), poly(vinylidene fluoride) (PVDF), cellulose acetate (CA), polysulfone (PSf), and polyethersulfone (PES) have been extensively explored to develop efficient UF membranes [51].

### 3.4.3 Coagulation

Wastewater treatment relies on coagulation to remove contaminants [54]. The principal physicochemical treatment for industrial effluent is coagulation–flocculation to reduce colloidal turbidity and suspended matter (Fig. 3.4) [55]. Coagulation can remove 40% of organic compounds and nitrogen from wastewater [56]. Pollutants and suspended matter collide with opposing particles to produce an insoluble agglomeration complex in sediment [57]. Several studies suggest that pH, coagulant dosage, and settling period impact wastewater contaminant removal. JAR test circumstances will be optimal because of these considerations. Typically, the coagulation/flocculation process consists of four steps: sweep coagulation, patch flocculation, charge neutralization, and bridging. The method of coagulation involves the presence of negatively charged colloidal particles, resulting in repulsion. The coagulant is used for stabilizing the colloid particulates, so neutralizing any repulsive force. A poly-electrolyte is usually added as a coagulant to almost zero the colloid's zeta potential. This is called the charge neutralization process [58]. A large amount of metal salts is added to the water, forming amorphous metallic hydroxides and large masses. This phenomenon is referred to as sweep coagulation [59]. The process of bridging flocculation can result in the formation of a quite large floating particles. This is mostly achieved by the adsorption of linear-chain molecules of high molecular weight, typically composed of polyacrylamide. This mechanism is simultaneous across several particles [60]. Meanwhile, the patch flocculation process happens as a result of the polymer being adsorbed onto the particles, causing a local charge reversal. An outcome of this procedure is the emergence of an attractive force between every individual particle [61].

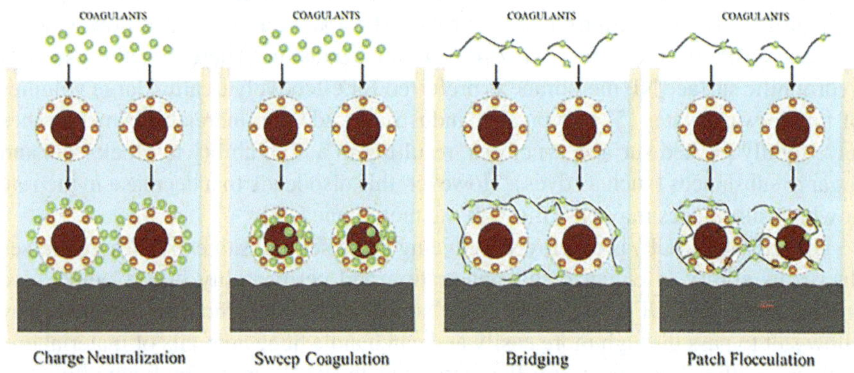


**Fig. 3.4** Coagulation method for wastewater treatment. Reproduced with permission from [62], Copyright 2021, AKJ journals

## 3.5 Chemical Methods

Several catalytic applications use chemical oxidation processes [63]. Advanced oxidation process (AOPs) is necessary for wastewater treatment. All wastewater treatment procedures that generate oxidizing species like hydroxyl radicals (˙OH) are called AOPs [64]. Electrochemical, photo-electrochemical, UV-assisted Fenton's oxidation, and ozonation are oxidation methods. During oxidation, catalysts and pH are crucial [65]. Figure 3.5 shows advanced textile wastewater oxidation techniques.

### 3.5.1 *Advanced Oxidation Processes*

AOPs were first proposed in the 1980s as a method for purifying drinking water [67, 68]. AOPs are characterized as oxidation processes that produce a significant amount of OH˙, which are capable of purifying water. Subsequently, the notion of AOP has been expanded to include oxidative reactions using sulfate radicals ($SO_4^{\cdot-}$). Unlike common oxidants such as Cl and $O_3$, which have dual roles as decontaminants and disinfectants, AOPs are primarily employed for the eradication of inorganic or organic contaminants in wastewater. Although the use of AOP to neutralize pathogens and pathogenic markers has been studied, it is rarely employed for disinfection since the radicals involved have an extremely brief half-life, measured in microseconds [69, 70]. As a result, the required duration for disinfection is not feasible because of the exceedingly low levels of radicals [71]. When AOPs are employed in the purification of polluted water, these extremely reactive radicals function as powerful oxidizing agents. It is expected that they will efficiently decompose pollutants in the contaminated water and transform them into less dangerous or non-toxic chemicals.

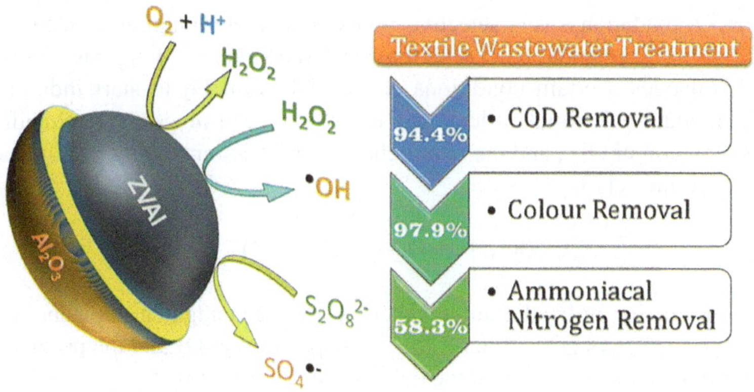


**Fig. 3.5** Water purification by AOPs. Reproduced with permission from [66], Copyright 2018, Elsevier

AOPs offer a promising and complete solution for the remediation of polluted water [72].

#### 3.5.1.1 Hydroxyl Radical-Based AOPs

The OH˙ is a highly reacted oxidizing agent in polluted water treatment. It possesses an oxidation potential ranging from 2.8 V (at pH 0) to 1.95 V (at pH 14) compared to the saturated calomel electrode (SCE), which is the generally used reference electrode [71]. The behavior of OH˙ is very indiscriminate, since it readily interacts with a wide range of species at very fast rates, often on the order of $10^8$–$10^{10}$ $M^{-1}$ $s^{-1}$. The OH˙ can target organic contaminants through four primary mechanisms: radical addition, radical abstraction, radical combination, and radical addition. When these compounds come into contact with organic compounds, they produce C-centered radicals, which can exist as either R· or R·–OH. Through the use of $O_2$, the C-centered radicals might undergo a transformation into organic ROO·. Every radical undergoes further reactions, leading to the production of very reactive species such as $H_2O_2$ and superoxide $O_2^{\cdot -}$. This process results in the chemical degradation and subsequent conversion of the organic compounds into minerals. OH˙, which have a brief existence, are generated on-site during application using several techniques. These procedures include a combination of oxidizing agents (including $O_3$ and $H_2O_2$), irradiation (e.g., ultrasonic and UV light), and catalysts (e.g., $Fe^{2+}$) [72]. The methods by which the principal AOPs generate hydroxyl radicals for wastewater treatment are outlined concisely here.

#### 3.5.1.2 Ozone-Based AOPs

A strong oxidizer, ozone ($O_3$) has a 2.07 V versus SCE oxidation potential. However, $O_3$ oxidation is very selective, with reaction rates ranging from $1.0 \times 10^0$–$10^3$ $M^{-1}$ $s^{-1}$. $O_3$ prefers to react with ionized and disintegrated organic compounds than neutral ones. Certain conditions create OH˙ from $O_3$ to start indirect non-selective oxidation. Various intricate processes have been suggested to elucidate the intricate creation of OH˙, and the comprehensive reaction including OH˙ generation is stated as follows [73]:

$$3O_3 + H_2O \rightarrow 2OH^{\cdot} + 4O_2 \tag{3.1}$$

The introduction of irradiation or other oxidants can augment the generation of OH. The $O_3/H_2O_2$ system augments the decomposition of $O_3$ and the production of OH˙ by producing $HO_2^-$ through the decomposition of $H_2O_2$.

$$H_2O_2 \rightarrow HO_2^- + H^+ \tag{3.2}$$

$$HO_2^- + O_3 \rightarrow OH^{\cdot} + O_2^- + O_2 \quad (3.3)$$

$O_3$ photolysis generates $H_2O_2$ as an extra oxidant in $O_3$/UV irradiation (Eq. 3.4)

$$O_3 + H_2O + hv \rightarrow H_2O_2 + O_2 \quad (3.4)$$

$$H_2O_2 + hv \rightarrow 2OH^{\cdot} \quad (3.5)$$

#### 3.5.1.3 UV-Based AOPs

Photons and catalysts or oxidants can generate hydroxyl radicals. RO-type SC $TiO_2$ is the most popular catalyst. Exciting $TiO_2$ particles produce $hv^+$ in the VB (oxidative capacity) and $e^-{}_{cb}$ in the CB (reductive capacity).

$$TiO_2 + hv \rightarrow e_{cb}^- + hv_{vb}^+ \quad (3.6)$$

When $OH^-$, $H_2O$, and $O_2{}^{\cdot -}$ react on $TiO_2$, $h^+$, and $e^-$ can produce $OH^{\cdot}$ [74].

$$hv_{vb}^+ + OH^-_{(surface)} \rightarrow OH^{\cdot} \quad (3.7)$$

$$hv_{vb}^+ + H_2O \rightarrow OH^{\cdot} + H^+ \quad (3.8)$$

$$e_{cb}^- + O_{2(absorbed)} \rightarrow O_2^{\cdot -} \quad (3.9)$$

The presence of $OH^-$, $H_2O$, and $O_2{}^{\cdot -}$ at the surface of $TiO_2$ leads to the production of $OH^{\cdot}$ through the interaction of these species with $h^+$ and $e^-$ [74].

$$H_2O_2 + hv \rightarrow 2OH^{\cdot} \quad (3.10)$$

$OH^{\cdot}$ may also be produced by photolysis of $H_2O$ below 242 nm.

$$H_2O_2 + hv \rightarrow OH^{\cdot} + H^{\cdot} \quad (3.11)$$

#### 3.5.1.4 Fenton-Related AOPs

Iron is the metal that is most often used because it can activate $H_2O_2$ and make $OH^{\cdot}$ in $H_2O$. The Fenton process involves the reaction of $H_2O_2$ with ferrous ions ($Fe^{2+}$) to produce highly reactive species. The reactive species produced are commonly recognized as $OH^{\cdot}$, however other molecules such ferryl ions have been proposed.

Prior research has already examined the extensive study of Fenton-related chemistry for the purpose of contaminated water treatment [75]. The subsequent reactions dominate conventional Fenton radical procedures [76]:

$$Fe^{2+} + H_2O_2 \rightarrow Fe^{3+} + OH^{\cdot} + OH^- \quad (3.12)$$

$$Fe^{3+} + H_2O_2 \rightarrow Fe^{2+} + HO_2^{\cdot} + H^+ \quad (3.13)$$

$$OH^{\cdot} + H_2O_2 \rightarrow HO_2^{\cdot} + H_2O \quad (3.14)$$

$$OH^{\cdot} + Fe^{2+} \rightarrow Fe^{3+} + OH^- \quad (3.15)$$

$$Fe^{3+} + HO_2^{\cdot} \rightarrow Fe^{2+} + O_2H^+ \quad (3.16)$$

$$Fe^{2+} + HO_2^{\cdot} + H^+ \rightarrow Fe^{3+} + H_2O_2 \quad (3.17)$$

$$2HO_2^{\cdot} \rightarrow H_2O_2 + O_2 \quad (3.18)$$

## 3.6 Biological Method

Biological treatment is the utilization of microorganisms to decompose the organic substances present in water and generate inorganic byproducts. The microorganisms are subsequently eliminated by the sedimentation technique. The process of wastewater purification, through aerobic and anaerobic processes, is a component of the biological therapy. The first growth process is the active growth process, sometimes referred to as the fixed film process. It encompasses several systems such as the intermittent filter, rotating biological contactor, trickling bed, and contact bed. This approach includes the conversion of biological materials by attaching it to an inert media, such as ceramic, rock, slag, or plastic [77]. Additionally, suspended growth techniques encompass aerated lagoons, activated oxidation ponds, and sludge processes. The techniques include either artificially or naturally suspending the microbial cells to transform organic waste [78]. Consequently, the residual suspended and organic particles resulting from the physical process are removed during the biological process, thereby achieving a water quality suitable for chemical treatment.

## 3.6.1 *Aerobic Processes*

Aerobic decomposition occurs in biodegradable organic matter when there is a enough supply of air or oxygen in the contaminated water, facilitated by aerobic and facultative microbes [79]. The extent of the process is depending upon the quantity of oxygen available, the duration of substance retention, the temperature, and the bacterial-driven biological activity. Moreover, the incorporation of certain compounds essential for bacterial growth might accelerate the biological oxidation process, facilitating the removal of organic pollutants from the surroundings. The process efficiently removes biological oxygen demand (BOD), nitrates, dissolved phosphates, chemical oxygen demand (COD), suspended organics, volatile organics, and other contaminants. This technique can achieve a decrease of up to 90% in the quantity of biodegradable organics present. One of the drawbacks of employing this procedure is the fabrication of a significant quantity of bio-solids, which necessitates extra, costlier treatment and management. Trickling filters, activated sludge processes, and oxidation ponds are the three technologies employed to accomplish the aerobic process.

### 3.6.1.1 Activated Sludge

Following initial sedimentation, the wastewater undergoes a biological process called "activated sludge." Despite early treatment, certain colloidal and suspended particles remain in wastewater, which are stirred up by air. The suspended particles gradually coalesce into bigger particles, while the colloidal particles provide a substrate for the growth of living organisms [80]. Activated sludge is a brownish material resembling foc (fraction of organic C in primary solids) and consists of organic matter produced from wastewater. This substance contains a wide range of microorganisms and other types of living organisms. Activated sludge is comprised of live microorganisms that has the capacity to assimilate or bind both colloidal and soluble organic substances. This is among the several advantageous characteristics of activated sludge. Living organisms metabolize the substance they absorb as sustenance and convert it into inactive, undissolved solids, as well as new microbe cells. The majority of this transition is conducted through a series of sequential steps. Specific bacteria metabolize complex compounds, converting them into simpler chemicals that are excreted as waste products [81]. Subsequently, microorganisms utilize the waste material to generate more basic chemicals. This procedure continues until the bacteria are no longer able to rely on the final waste products. To prevent the activated sludge from settling at the bottom of the tank as it interacts with the polluted water being treated, a kind of agitation must be employed. This agitation will sustain the suspension of the activated sludge.

#### 3.6.1.2 Trickling Filter

A trickling filter is an instance of an aerobic treatment method. Through this procedure, the microorganisms employed for remediation attach to a substrate in order to eliminate organic waste from polluted water. Microorganisms may adhere to many substances like rock, sand, redwood, and even synthetic materials. The microorganisms that colonize the filtering media are commonly known as "biological slime," encompassing bacteria, fungi, and algae. As part of the procedure, the polluted water is dispersed and subsequently let to percolate through the medium. During this period, the bacterial slime attached to the medium will metabolize the existing organic matter in the wastewater. The thickness of the biological slime rises when the organic material present in contaminated water is transformed into new cellular material. Over time, the biological slime layer will accumulate and finally detach from the surface. Sloughing refers to the phenomenon when the slime layer detaches and separates. The detached slime layer, which contains particles and liquids, is collected via under drainage together with the treated wastewater [82].

#### 3.6.1.3 Anaerobic Processes

Anaerobic decomposition, commonly referred to as putrefaction, happens when there is an inadequate amount of dissolved oxygen in the wastewater. Anaerobic and facultative bacteria decompose the complex chemical compound into simpler organic molecules including sulfur, C, and nitrogen. Hydrogen sulfide, ammonia, methane, and nitrogen are the four primary gases generated as a result of this process. This approach is employed to reduce the biological load carried by wastewater [69, 83–88].

## 3.7 Influencing Factors for Wastewater Treatment

The subsequent sections outline the impact of temperature, pH, and concentration of natural organic matter, and ionic strength on the elimination of heavy metals from residential wastewater effluents, lakes, rivers, and groundwater.

### *3.7.1 Temperature*

Tests indicate that higher temperatures improve elimination effectiveness across several treatments [89]. Plants were more successful in removing $Cd^{2+}$, $Cu^{2+}$, and $Ni^{2+}$ from the environment when the temperature reached 55 °C [90]. Researchers examined the impact of temperature on the adsorption of copper ($Cu^{2+}$) and lead ($Pb^{2+}$) ions in hydrogels at four distinct temperatures (22, 30, 35, and 40 °C). $Pb^{2+}$

optimum temperature was 30 °C, while $Cu^{2+}$ optimum temperature was 35 °C. A temperature rise of up to 40 °C improved removal results [91]. Increasing the temperature from 5 to 40 °C improved the efficiency of pistachio residues in removing $Cr^{6+}$. This phenomenon was linked to more adsorption sites on the material surface [92]. As temperature climbed from 10 to 60 °C in a research using fungus biomass, heavy metal biosorption efficacy increased. This is due to the biomass substrates' active sites having a higher affinity [93]. However, even little temperature changes may significantly affect heavy metal elimination. Using magnetic iron oxide NPs as adsorbents lowered $Ni^{2+}$, $Cu^{2+}$, and $Al^{3+}$ removal effectiveness with increasing temperature. The chemisorption rate at higher temperatures caused heavy metals to not reach new active sites on the particle surface, reducing their ability to remove metals. Moreover, heating can change NPs pores, reducing adsorption capability [94].

### 3.7.2 Ionic Strength

Ionic strength is determined by the concentration and chemical charges of ions in an aqueous solution. This strength impacts heavy metal removal and ion exchange between metal ions and adsorbents. Increased sodium chloride (NaCl) concentration reduced ion adsorption on porous magnetic particles, including $Cr^{3+}$, $Zn^{2+}$, and $Cu^{2+}$ [95]. During the investigation, it was observed that the sorption of $Cu^{2+}$ and $Ni^{2+}$ on bentonite increased as the ionic strength decreased at pH 5.5. The highest sorption was found at NaCl concentrations below 0.05 M [96]. Increased ionic strength may reduce electrostatic repulsion, leading to particle aggregation and reduced binding sites and sorption. The presence of chloride may help in the development of difficult-to-remove soluble metal chloride complexes from water [97]. This is because chloride compounds have a poor adsorption affinity [98].

### 3.7.3 Natural Organic Matter

Biosphere–geosphere interactions like plant and animal breakdown in water produce natural organic matter (NOM) in various water sources. However, it is more commonly found in surface waters. Consequently, NOM has the ability to undergo chemical interactions with heavy metals present in contaminated water through several mechanisms such as adsorption, chelation, and ion exchange. NOM impacts heavy metal reactivity, mobility, bioavailability, and toxicity, but its impact on heavy metal elimination is difficult to quantify. Lead can form complexes with naturally present microbes in natural streams or wastewater, resulting in increased mobility and resistance to removal using conventional procedures such as coagulation/flocculation and chemical precipitation. A research using titanate nanotubes supported by powder-activated charcoal showed that complexes and weak interactions between titanate nanotubes and dissolved organic matter affect $Pb^{2+}$ adsorption [99]. High quantities of NOM can hinder heavy metal absorption and removal due to complex formation, chelation, and ionic competition [100]. The existence of these characteristics and their interaction pose substantial difficulties for the elimination of heavy metals from polluted water. Furthermore, considering their potential impact on the ability to replicate findings, it is essential to consider these factors while evaluating different methods of elimination.

### *3.7.4 pH*

Fluctuations in pH can affect the removal of existing pollutants by altering the equilibrium between inorganic and organic chemicals in water. The modifications might arise either from the introduction of chemicals during the restoration procedure or from inherent biological instability [101]. Heavy metals become cationic and exhibit increased mobility and solubility in water when the pH value is neutral or lower. Elevated pH values, however, impact their oxidation state and promote the formation of hydroxides and other anions [102]. For example, the pH of the solution has an impact on both the effectiveness of removing natural clay and its ability to adsorb $Ni^{2+}$ when used for elimination purposes. The efficiency of nickel removal increased in direct proportion to the pH of the solution [103].

## References

1. Sharma, D., Kansal, A.: Water quality analysis of River Yamuna using water quality index in the national capital territory, India (2000–2009). Appl. Water Sci. **1**, 147–157 (2011). https://doi.org/10.1007/s13201-011-0011-4
2. Barton, A.: Water in crisis—Sudan. Water Proj
3. Bănăduc, D., Simić, V., Cianfaglione, K., et al.: Freshwater as a sustainable resource and generator of secondary resources in the 21st century: stressors, threats, risks, management and protection strategies, and conservation approaches. Int. J. Environ. Res. Public Health **19**, 16570 (2022)
4. du Plessis, A.: Primary water quality challenges, contaminants and the world's dirtiest places. In: Springer Water, pp. 79–114. Springer Nature (2019)
5. Boretti, A.: Rosa, L.: Reassessing the projections of the world water development report. NPJ Clean Water **2** (2019). https://doi.org/10.1038/s41545-019-0039-9
6. Perveen, S., Amar-Ul-Haque: Drinking water quality monitoring, assessment and management in Pakistan: a review. Heliyon **9** (2023)
7. Allaq, A.A.A., Mahid, H.H., Yahya, E.B., et al.: Emerging drinking water borne diseases: a review on types, sources and health precaution. J. Pharm. Res. Int. **35**, 1–17 (2023). https://doi.org/10.9734/jpri/2023/v35i317462
8. Wato, T., Amare, M., Bonga, E., et al.: The agricultural water pollution and its minimization strategies—a review. J. Resour. Dev. Manag. **64**, 2020 (2020). https://doi.org/10.7176/jrdm/64-02
9. Zahoor, I., Mushtaq, A.: Water pollution from agricultural activities: a critical global review. Int. J. Chem. Biochem. Sci. **23**, 164–176 (2023)

10. Saquib, S., Yadav, A.K., Prajapati, K.B.: Emerging pollutants in water and human health. In: Contamination of Water: Health Risk Assessment and Treatment Strategies, pp. 285–299 (2021)
11. Hunt, C.F., Voulvoulis, N.: Chemical pollution of the aquatic environment and health. Issues Environ. Sci. Technol. 39–69 (2021). https://doi.org/10.1039/9781839160431-00039
12. Carrard, N., Foster, T., Willetts, J.: Groundwater as a source of drinking water in southeast Asia and the Pacific: a multi-country review of current reliance and resource concerns. Water (Switzerland) **11** (2019). https://doi.org/10.3390/w11081605
13. Chowdhury, S., Balasubramanian, R.: Recent advances in the use of graphene-family nanoadsorbents for removal of toxic pollutants from wastewater. Adv. Colloid Interface Sci. **204**, 35–56 (2014)
14. Chowdhary, P., Bharagava, R.N., Mishra, S., Khan, N.: Role of industries in water scarcity and its adverse effects on environment and human health. In: Environmental Concerns and Sustainable Development, pp. 235–256. Springer Singapore (2020)
15. Chen, B., Wang, M., Duan, M., et al.: In search of key: Protecting human health and the ecosystem from water pollution in China. J. Clean. Prod. **228**, 101–111 (2019). https://doi.org/10.1016/j.jclepro.2019.04.228
16. Wu, H., Gai, Z., Guo, Y., et al.: Does environmental pollution inhibit urbanization in China? A new perspective through residents' medical and health costs. Environ. Res. **182** (2020). https://doi.org/10.1016/j.envres.2020.109128
17. Parris, K.: Impact of agriculture on water pollution in OECD countries: recent trends and future prospects. Int. J. Water Resour. Dev. **27**, 33–52 (2011). https://doi.org/10.1080/07900627.2010.531898
18. Moss, B.: Water pollution by agriculture. Philos. Trans. R. Soc. B Biol. Sci. **363**, 659–666 (2008)
19. Lai, W.: Pesticide use and health outcomes: Evidence from agricultural water pollution in China. J. Environ. Econ. Manage. **86**, 93–120 (2017). https://doi.org/10.1016/j.jeem.2017.05.006
20. Xiao, J., Wang, L., Deng, L., Jin, Z.: Characteristics, sources, water quality and health risk assessment of trace elements in river water and well water in the Chinese Loess Plateau. Sci. Total. Environ. **650**, 2004–2012 (2019). https://doi.org/10.1016/j.scitotenv.2018.09.322
21. Wu, C., Maurer, C., Wang, Y., et al.: Water pollution and human health in China. Environ. Health Perspect. **107**, 251–256 (1999)
22. Ustaoğlu, F., Tepe, Y., Taş, B.: Assessment of stream quality and health risk in a subtropical Turkey river system: a combined approach using statistical analysis and water quality index. Ecol. Ind. **113** (2020). https://doi.org/10.1016/j.ecolind.2019.105815
23. Halder, J., Islam, N.: Water pollution and its impact on the human health. J. Environ. Hum. **2**, 36–46 (2015). https://doi.org/10.15764/eh.2015.01005
24. Ahmed, S.N., Haider, W.: Heterogeneous photocatalysis and its potential applications in water and wastewater treatment: a review. Nanotechnology **29** (2018)
25. Nguyen, C.H., Juang, R.S.: Efficient removal of cationic dyes from water by a combined adsorption-photocatalysis process using platinum-doped titanate nanomaterials. J. Taiwan Inst. Chem. Eng. **99**, 166–179 (2019). https://doi.org/10.1016/j.jtice.2019.03.017
26. Wawrzkiewicz, M., Polska-Adach, E., Hubicki, Z.: Application of titania based adsorbent for removal of acid, reactive and direct dyes from textile effluents. Adsorption (2019). https://doi.org/10.1007/s10450-019-00062-0
27. Wong, J.K.H., Tan, H.K., Lau, S.Y., et al.: Potential and challenges of enzyme incorporated nanotechnology in dye wastewater treatment: a review. J. Environ. Chem. Eng. **7** (2019)
28. Yahya, N., Aziz, F., Jamaludin, N.A., et al.: A review of integrated photocatalyst adsorbents for wastewater treatment. J. Environ. Chem. Eng. **6**, 7411–7425 (2018). https://doi.org/10.1016/j.jece.2018.06.051
29. Rashid, R., Shafiq, I., Akhter, P., et al.: A state-of-the-art review on wastewater treatment techniques: the effectiveness of adsorption method. Environ. Sci. Pollut. Res. **28**, 9050–9066 (2021)

30. Samsami, S., Mohamadi, M., Sarrafzadeh, M.H., et al.: Recent advances in the treatment of dye-containing wastewater from textile industries: overview and perspectives. Process. Saf. Environ. Prot. **143**, 138–163 (2020)
31. Foroutan, R., Mohammadi, R., Farjadfard, S., et al.: Characteristics and performance of Cd, Ni, and Pb bio-adsorption using *Callinectes sapidus* biomass: real wastewater treatment. Environ. Sci. Pollut. Res. **26**, 6336–6347 (2019). https://doi.org/10.1007/s11356-018-04108-8
32. Hu, H., Xu, K.: Physicochemical technologies for HRPs and risk control. In: High-Risk Pollutants in Wastewater, pp. 169–207. Elsevier (2019)
33. Tara, N., Siddiqui, S.I., Rathi, G., et al.: Nano-engineered adsorbent for the removal of dyes from water: a review. Curr. Anal. Chem. **16**, 14–40 (2019). https://doi.org/10.2174/1573411015666190117124344
34. Carvalho, J., Araujo, J., Castro, F.: Alternative low-cost adsorbent for water and wastewater decontamination derived from eggshell waste: an overview. Waste Biomass Valorization **2**, 157–167 (2011)
35. Sanghi, R., Bhattacharya, B.: Review on decolorisation of aqueous dye solutions by low cost adsorbents. Color. Technol. **118**, 256–269 (2002). https://doi.org/10.1111/j.1478-4408.2002.tb00109.x
36. Jadhav, A.J., Srivastava, V.C.: Adsorbed solution theory based modeling of binary adsorption of nitrobenzene, aniline and phenol onto granulated activated carbon. Chem. Eng. J. **229**, 450–459 (2013). https://doi.org/10.1016/j.cej.2013.06.021
37. Galán, J., Rodríguez, A., Gómez, J.M., et al.: Reactive dye adsorption onto a novel mesoporous carbon. Chem. Eng. J. **219**, 62–68 (2013). https://doi.org/10.1016/j.cej.2012.12.073
38. Gupta, V.K., Gupta, B., Rastogi, A., et al.: A comparative investigation on adsorption performances of mesoporous activated carbon prepared from waste rubber tire and activated carbon for a hazardous azo dye-Acid Blue 113. J. Hazard. Mater. **186**, 891–901 (2011). https://doi.org/10.1016/j.jhazmat.2010.11.091
39. Gupta, N., Kushwaha, A.K., Chattopadhyaya, M.C.: Application of potato (*Solanum tuberosum*) plant wastes for the removal of methylene blue and malachite green dye from aqueous solution. Arab. J. Chem. **9**, S707–S716 (2016). https://doi.org/10.1016/j.arabjc.2011.07.021
40. Al-Amrani, W.A., Hanafiah, M.A.K.M., Mohammed, A.H.A.: A comprehensive review of anionic azo dyes adsorption on surface-functionalised silicas. Environ. Sci. Pollut. Res. **29**, 76565–76610 (2022)
41. Sumanjit, R.S., Mahajan, R.K.: Equilibrium, kinetics and thermodynamic parameters for adsorptive removal of dye Basic Blue 9 by ground nut shells and Eichhornia. Arab. J. Chem. **9**, S1464–S1477 (2016). https://doi.org/10.1016/j.arabjc.2012.03.013
42. Auta, M., Hameed, B.H.: Preparation of waste tea activated carbon using potassium acetate as an activating agent for adsorption of Acid Blue 25 dye. Chem. Eng. J. **171**, 502–509 (2011). https://doi.org/10.1016/j.cej.2011.04.017
43. Kumar, R., Ahmad, R.: Biosorption of hazardous crystal violet dye from aqueous solution onto treated ginger waste (TGW). Desalination **265**, 112–118 (2011). https://doi.org/10.1016/j.desal.2010.07.040
44. Zhang, W., Dong, L., Yan, H., et al.: Removal of methylene blue from aqueous solutions by straw based adsorbent in a fixed-bed column. Chem. Eng. J. **173**, 429–436 (2011). https://doi.org/10.1016/j.cej.2011.08.001
45. Belala, Z., Jeguirim, M., Belhachemi, M., et al.: Biosorption of basic dye from aqueous solutions by date stones and palm-trees waste: kinetic, equilibrium and thermodynamic studies. Desalination **271**, 80–87 (2011). https://doi.org/10.1016/j.desal.2010.12.009
46. Tunali Akar, S., Gorgulu, A., Akar, T., Celik, S.: Decolorization of Reactive Blue 49 contaminated solutions by *Capsicum annuum* seeds: batch and continuous mode biosorption applications. Chem. Eng. J. **168**, 125–133 (2011). https://doi.org/10.1016/j.cej.2010.12.049
47. Moussavi, G., Khosravi, R.: The removal of cationic dyes from aqueous solutions by adsorption onto pistachio hull waste. Chem. Eng. Res. Des. **89**, 2182–2189 (2011). https://doi.org/10.1016/j.cherd.2010.11.024

48. Kara, I., Akar, S.T., Akar, T., Ozcan, A.: Dithiocarbamated *Symphoricarpus albus* as a potential biosorbent for a reactive dye. Chem. Eng. J. **211–212**, 442–452 (2012). https://doi.org/10.1016/j.cej.2012.09.086
49. Zhong, Q.Q., Yue, Q.Y., Li, Q., et al.: Preparation, characterization of modified wheat residue and its utilization for the anionic dye removal. Desalination **267**, 193–200 (2011). https://doi.org/10.1016/j.desal.2010.09.025
50. Elkady, M.F., Ibrahim, A.M., El-Latif, M.M.A.: Assessment of the adsorption kinetics, equilibrium and thermodynamic for the potential removal of reactive red dye using eggshell biocomposite beads. Desalination **278**, 412–423 (2011). https://doi.org/10.1016/j.desal.2011.05.063
51. Yang, C., Xu, W., Nan, Y., et al.: Fabrication and characterization of a high performance polyimide ultrafiltration membrane for dye removal. J. Colloid Interface Sci. **562**, 589–597 (2020). https://doi.org/10.1016/j.jcis.2019.11.075
52. Zhu, Z., Liu, D., Cai, S., et al.: Dyes Removal by Composite Membrane of Sepiolite Impregnated Polysulfone Coated by Chemical Deposition of Tea Polyphenols. Elsevier (2020)
53. Ye, W., Ye, K., Lin, F., et al.: Enhanced fractionation of dye/salt mixtures by tight ultrafiltration membranes via fast bio-inspired co-deposition for sustainable textile wastewater management. Chem. Eng. J. **379** (2020). https://doi.org/10.1016/j.cej.2019.122321
54. Ang, W.L., Mohammad. A.W.: State of the art and sustainability of natural coagulants in water and wastewater treatment. J. Clean. Prod. **262** (2020)
55. Gautam, S., Saini, G.: Use of natural coagulants for industrial wastewater treatment. Glob. J. Environ. Sci. Manag. **6**, 553–578 (2020). https://doi.org/10.22034/gjesm.2020.04.10
56. Loloei, M., Nekonam, G., Alidadi, H., Kor, Y.: Study of the coagulation process in wastewater treatment of dairy industries. Int. J. Environ. Health Eng. **3**, 12 (2014). https://doi.org/10.4103/2277-9183.132684
57. Syam Babu, D., Anantha Singh, T.S., Nidheesh, P.V., Suresh Kumar, M.: Industrial wastewater treatment by electrocoagulation process. Sep. Sci. Technol. **55**, 3195–3227 (2020)
58. Amran, A.H., Zaidi, N.S., Muda, K., Loan, L.W.: Effectiveness of natural coagulant in coagulation process: a review. Int. J. Eng. Technol. **7**, 34–37 (2018). https://doi.org/10.14419/ijet.v7i3.9.15269
59. Zhao, Y.X., Gao, B.Y., Shon, H.K., et al.: The effect of second coagulant dose on the regrowth of flocs formed by charge neutralization and sweep coagulation using titanium tetrachloride ($TiCl_4$). J. Hazard. Mater. **198**, 70–77 (2011). https://doi.org/10.1016/j.jhazmat.2011.10.015
60. Hogg, R.: Bridging flocculation by polymers. KONA Powder Part. J. **30**, 3–14 (2012)
61. Hjorth, M., Jørgensen, B.U.: Polymer flocculation mechanism in animal slurry established by charge neutralization. Water Res. **46**, 1045–1051 (2012). https://doi.org/10.1016/j.watres.2011.11.078
62. Sukmana, H., Bellahsen, N., Pantoja, F., Hodur, C.: Adsorption and coagulation in wastewater treatment—review. Prog. Agric. Eng. Sci. **17**, 49–68 (2021). https://doi.org/10.1556/446.2021.00029
63. Shafiq, I., Hussain, M., Shehzad, N., et al.: The effect of crystal facets and induced porosity on the performance of monoclinic $BiVO_4$ for the enhanced visible-light driven photocatalytic abatement of methylene blue. J. Environ. Chem. Eng. **7** (2019). https://doi.org/10.1016/j.jece.2019.103265
64. Grande, G.A.: Treatment of wastewater from textile dyeing by ozonization (2015)
65. Ahmad, A., Mohd-Setapar, S.H., Chuong, C.S., et al.: Recent advances in new generation dye removal technologies: novel search for approaches to reprocess wastewater. RSC Adv. **5**, 30801–30818 (2015)
66. Khatri, J., Nidheesh, P.V., Anantha Singh, T.S., Suresh Kumar, M.: Advanced oxidation processes based on zero-valent aluminium for treating textile wastewater. Chem. Eng. J. **348**, 67–73 (2018). https://doi.org/10.1016/j.cej.2018.04.074
67. Glaze, W.H., Kang, J.W., Chapin, D.H.: The chemistry of water treatment processes involving ozone, hydrogen peroxide and ultraviolet radiation. Ozone Sci. Eng. **9**, 335–352 (1987). https://doi.org/10.1080/01919518708552148

68. Glaze, W.H.: Drinking-water treatment with ozone. Environ. Sci. Technol. **21**, 224–230 (1987). https://doi.org/10.1021/es00157a001
69. Ikai, H., Nakamura, K., Shirato, M., et al.: Photolysis of hydrogen peroxide, an effective disinfection system via hydroxyl radical formation. Antimicrob. Agents Chemother. **54**, 5086–5091 (2010). https://doi.org/10.1128/AAC.00751-10
70. Cho, M., Chung, H., Choi, W., Yoon, J.: Different inactivation behaviors of MS-2 phage and *Escherichia coli* in $TiO_2$ photocatalytic disinfection. Appl. Environ. Microbiol. **71**, 270–275 (2005). https://doi.org/10.1128/AEM.71.1.270-275.2005
71. Bixio, D., Wintgens, T., Ravazzini, A., et al.: Water reuse. In: Wastewater Quality Monitoring and Treatment, pp. 329–349 (2007)
72. Huang, C.P., Dong, C., Tang, Z.: Advanced chemical oxidation: Its present role and potential future in hazardous waste treatment. Waste Manag. **13**, 361–377 (1993). https://doi.org/10.1016/0956-053X(93)90070-D
73. Gottschalk, C., Libra, J.A., Saupe, A.: Ozonation of water and waste water: a practical guide to understanding ozone and its application (2008)
74. Tang, W.Z.: Physicochemical treatment of hazardous wastes (2016)
75. Pignatello, J.J., Oliveros, E., MacKay, A.: Advanced oxidation processes for organic contaminant destruction based on the Fenton reaction and related chemistry. Crit. Rev. Environ. Sci. Technol. **36**, 1–84 (2006)
76. Deng, Y., Zhao, R.: Advanced oxidation processes (AOPs) in wastewater treatment. Curr. Pollut. Rep **1**, 167–176 (2015). https://doi.org/10.1007/s40726-015-0015-z
77. Raju, S., Carbery, M., Kuttykattil, A., et al.: Improved methodology to determine the fate and transport of microplastics in a secondary wastewater treatment plant. Water Res. **173** (2020). https://doi.org/10.1016/j.watres.2020.115549
78. Kumar, R., Sarmah, A.K., Padhye, L.P.: Fate of pharmaceuticals and personal care products in a wastewater treatment plant with parallel secondary wastewater treatment train. J. Environ. Manage. **233**, 649–659 (2019). https://doi.org/10.1016/j.jenvman.2018.12.062
79. Barragán, B.E., Costa, C., Carmen Márquez, M.: Biodegradation of azo dyes by bacteria inoculated on solid media. Dye. Pigment. **75**, 73–81 (2007). https://doi.org/10.1016/j.dyepig.2006.05.014
80. Zhang, H., Feng, J., Chen, S., et al.: Geographical patterns of nirS gene abundance and nirS-type denitrifying bacterial community associated with activated sludge from different wastewater treatment plants. Microb. Ecol. **77**, 304–316 (2019). https://doi.org/10.1007/s00248-018-1236-7
81. Sepehri, A., Sarrafzadeh, M.H., Avateffazeli, M.: Interaction between *Chlorella vulgaris* and nitrifying-enriched activated sludge in the treatment of wastewater with low C/N ratio. J. Clean Prod. **247** (2020). https://doi.org/10.1016/j.jclepro.2019.119164
82. Thatai, S., Verma, R., Khurana, P., et al.: Water quality standards, its pollution and treatment methods. In: A New Generation Material Graphene: Applications in Water Technology, pp. 21–42. Springer International Publishing (2018)
83. Bernard, O., Hadj-Sadok, Z., Dochain, D., et al.: Dynamical model development and parameter identification for an anaerobic wastewater treatment process. Biotechnol. Bioeng. **75**, 424–438 (2001). https://doi.org/10.1002/bit.10036
84. Bernet, N., Delgenes, N., Akunna, J.C., et al.: Combined anaerobic-aerobic SBR for the treatment of piggery wastewater. Water Res. **34**, 611–619 (2000). https://doi.org/10.1016/S0043-1354(99)00170-0
85. Fux, C., Boehler, M., Huber, P., et al.: Biological treatment of ammonium-rich wastewater by partial nitritation and subsequent anaerobic ammonium oxidation (anammox) in a pilot plant. J. Biotechnol. **99**, 295–306 (2002). https://doi.org/10.1016/S0168-1656(02)00220-1
86. Talarposhti, A.M., Donnelly, T., Anderson, G.K.: Colour removal from a simulated dye wastewater using a two-phase anaerobic packed bed reactor. Water Res. **35**, 425–432 (2001). https://doi.org/10.1016/S0043-1354(00)00280-3
87. Van Der Zee, F.P., Villaverde, S.: Combined anaerobic-aerobic treatment of azo dyes—a short review of bioreactor studies. Water Res. **39**, 1425–1440 (2005)

88. Venkata Mohan, S., Lalit Babu, V., Sarma, P.N.: Anaerobic biohydrogen production from dairy wastewater treatment in sequencing batch reactor (AnSBR): effect of organic loading rate. Enzyme Microb. Technol. **41**, 506–515 (2007). https://doi.org/10.1016/j.enzmictec.2007.04.007
89. Chen, H., Zhao, J., Dai, G., et al.: Adsorption characteristics of Pb(II) from aqueous solution onto a natural biosorbent, fallen *Cinnamomum camphora* leaves. Desalination **262**, 174–182 (2010). https://doi.org/10.1016/j.desal.2010.06.006
90. Xu, X., Yang, Y., Wang, G., et al.: Removal of heavy metals from industrial sludge with new plant–based washing agents. Chemosphere **246** (2020). https://doi.org/10.1016/j.chemosphere.2020.125816
91. Tirtom, V.N., Dinçer, A.: Effective removal of heavy metals from an aqueous solution with poly(*N*-vinylimidazole-acrylamide) hydrogels. Sep. Sci. Technol. **56**, 912–924 (2021). https://doi.org/10.1080/01496395.2020.1735434
92. Moussavi, G., Barikbin, B.: Biosorption of chromium(VI) from industrial wastewater onto pistachio hull waste biomass. Chem. Eng. J. **162**, 893–900 (2010). https://doi.org/10.1016/j.cej.2010.06.032
93. Alothman, Z.A., Bahkali, A.H., Khiyami, M.A., et al.: Low cost biosorbents from fungi for heavy metals removal from wastewater. Sep. Sci. Technol. **55**, 1766–1775 (2020). https://doi.org/10.1080/01496395.2019.1608242
94. Anbalagan, K., Kumar, M.M., Sudarsan, J.S., Nithiyanantham, S.: Removal of heavy metal ions from industrial wastewater using magnetic nanoparticles. J. Eng. Res. **10**, 59–71 (2022). https://doi.org/10.36909/jer.6924
95. Hu, X., Hu, Y., Xu, G., et al.: Green synthesis of a magnetic $β$-cyclodextrin polymer for rapid removal of organic micro-pollutants and heavy metals from dyeing wastewater. Environ. Res. **180** (2020). https://doi.org/10.1016/j.envres.2019.108796
96. Musso, T.B., Parolo, M.E., Pettinari, G.: pH, ionic strength, and ion competition effect on Cu(II) and Ni(II) sorption by a Na-bentonite used as liner material. Polish. J. Environ. Stud. **28**, 2299–2310 (2019). https://doi.org/10.15244/pjoes/84922
97. Joseph, L., Jun, B.M., Flora, J.R.V., et al.: Removal of heavy metals from water sources in the developing world using low-cost materials: a review. Chemosphere **229**, 142–159 (2019)
98. Villarín, M.C., Merel, S.: Paradigm shifts and current challenges in wastewater management. J. Hazard Mater. **390** (2020). https://doi.org/10.1016/j.jhazmat.2020.122139
99. Ma, J., Li, F., Qian, T., et al.: Natural organic matter resistant powder activated charcoal supported titanate nanotubes for adsorption of Pb(II). Chem. Eng. J. **315**, 191–200 (2017). https://doi.org/10.1016/j.cej.2017.01.029
100. Lin, L., Xu, X., Papelis, C., Xu, P.: Innovative use of drinking water treatment solids for heavy metals removal from desalination concentrate: synergistic effect of salts and natural organic matter. Chem. Eng. Res. Des. **120**, 231–239 (2017). https://doi.org/10.1016/j.cherd.2017.02.009
101. Ncube, P., Pidou, M., Stephenson, T., et al.: Consequences of pH change on wastewater depth filtration using a multimedia filter. Water Res. **128**, 111–119 (2018). https://doi.org/10.1016/j.watres.2017.10.040
102. Taşar, Ş, Kaya, F., Özer, A.: Biosorption of lead(II) ions from aqueous solution by peanut shells: equilibrium, thermodynamic and kinetic studies. J. Environ. Chem. Eng. **2**, 1018–1026 (2014). https://doi.org/10.1016/j.jece.2014.03.015
103. Es-Sahbany, H., Berradi, M., Nkhili, S., et al.: Removal of heavy metals (nickel) contained in wastewater-models by the adsorption technique on natural clay. In: Materials Today: Proceedings, pp. 866–875 (2019)

# Chapter 4
# Photocatalytic Application

## 4.1 Photocatalysis

Photocatalyst (PC) is a combination of light (photon) and catalyst (substance that alters reaction rate). Thus, PC modifies chemical reaction rates when exposed to light. This phenomenon is called photocatalytic activity (PCA). PCA involves processes using light and an SC. PC absorbs light and catalyzes chemical processes. Each PC is fundamentally SC. PCA generates an $e^-$–$h^+$ pair in semiconducting materials exposed to light. Typically, "PCA" involves using sunlight to break down large molecules into smaller, less poisonous, or lighter species. This PCA technique uses light to oxidize or reduce organic contaminants through simple redox processes [1]. This approach can reduce energy usage by converting light energy into chemical energy for breaking down and removing refractory organic molecules [2]. PCA beats traditional methods due to its convenience, ability to work at ambient temperatures and pressures, total toxin degradation, cost-effectiveness, absence of secondary pollution, and ppm/ppb pollution reduction [3]. PCA is a sustainable approach for environmental cleaning, eliminating the need for traditional wastewater treatment methods. The PCA process may occur in both aqueous and air environments [4]. PCA can remove many organic contaminants, such as antibiotics, synthetic chemicals, polymers, persistent organic pollutants, organic solvents, ionic liquids, surfactants, and polycyclic aromatic hydrocarbons, in addition to dyes. Additionally, this technology is used to divide water and remove $CO_2$ for fuel and other important items [5, 6]. The biggest hurdles in PCA include extracting the catalyst from the slurry and activating the surface with ultraviolet light. PCA offers advantages over traditional technologies such as complete mineralization of contaminants, cleanliness, lack of contaminants, economics, operation at normal pressure and temperature, and the capability of lowering contaminants to concentrations as low as ppm and ppb [7, 8]. Furthermore, the technique employs solar and visible light as a cost-effective energy source [3]. The PCA can be treated using UV photolysis, photooxidation with $OH^{\cdot}$, $O_3$, and $H_2O_2$ [9]. Furthermore, PCA obviates the need for traditional techniques of

 
M. Ikram et al., *Carbon-Based Nanomaterials for Environmental Applications*,
Engineering Materials, https://doi.org/10.1007/978-3-031-59390-1_4

treating wastewater, since it is an environmentally friendly technology in the realm of purifying the environment [7, 10, 11]. Figure 4.1 enumerates some distinctive benefits of PCA. Another crucial characteristic of PCA is its ability to perform this procedure in the atmosphere and in a water-based solution [4, 12–14]. Besides dyes, the PCA technique is employed for a range of other organic contaminants such as polymers, antibiotics, organic solvents, insecticides, ionic liquids, and polycyclic aromatic hydrocarbons [6, 15–25].

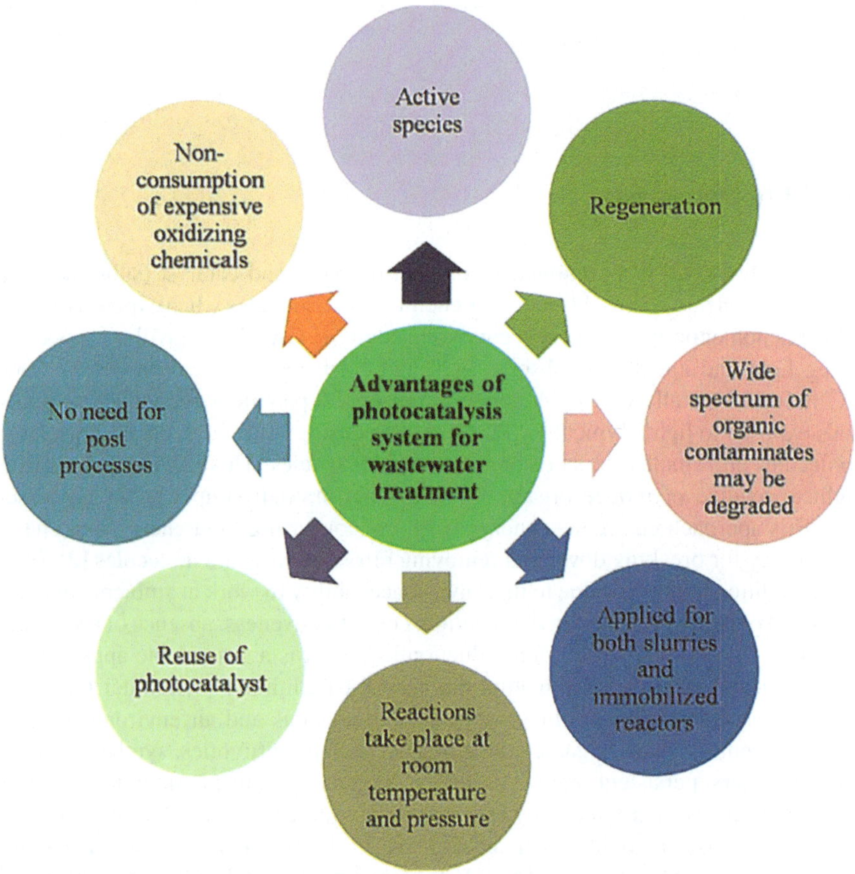


**Fig. 4.1** Advantages of PCA for water purification. Reproduced with permission from [26], copyright 2020, Elsevier

## 4.2 Various Mechanisms of Photocatalysis

### *4.2.1 Conventional Mechanism*

Heterogeneous PCA occurs on SC PC when exposed to light [27]. A PC generates $h^+$ in the VB by stimulating $e^-$ in the SC from the VB to the CB when subjected to light with photon energy larger than its Eg. Photoexcited $e^-$–$h^+$ pairs can recombine, generating heat energy that hinders photocatalysis, or travel to the PC surface and interact with adsorbed molecules. $O_2$ molecules capture the $e^-$ of CB to make $^{\cdot}O_2^-$, whereas $h^+$ reacts with $H_2O$ to form $OH^{\cdot}$ (oxidative potential +2.8 V). These highly reactive radicals efficiently and indiscriminately break down contaminants into less harmful substances. The potential interfacial reactions may be expressed using the following equations from 4.1 to 4.5 (Fig. 4.2) [28–34]:

$$\text{Photocatalyst} + \text{hv(UV)} \rightarrow e_{cb}^- + h_{vb}^+ \tag{4.1}$$

$$O_2 + e_{cb}^- \rightarrow {}^{\cdot}O_2^- \tag{4.2}$$

$$h_{vb}^+ + H_2O \rightarrow {}^{\cdot}OH \tag{4.3}$$

$$\text{Dye} + h_{vb}^+ \rightarrow \text{Oxidation product} \tag{4.4}$$

$$\text{Dye} + e_{cb}^- \rightarrow \text{Reduction product} \tag{4.5}$$

### *4.2.2 Advanced Mechanism*

The band orientation among two SC materials' CBs and VBs divides heterojunctions into three kinds. These types are: Type I (straddling-gap), Type II (staggered-gap), and Type III (broken-gap junctions), as revealed in Fig. 4.3a–c. Type II accumulate $e^-$ and $h^+$ in the SC with a narrower Eg. Consequently, there is a significant likelihood of recombination, leading to subpar efficacy of PCA. Type II junction allows $e^-$ and $h^+$ to transfer to lower energy areas, with $e^-$ concentrating in photosystem II (PS II) and $h^+$ in PS I. Consequently, the PCA will increase because the charge carriers are physically separated. Due to $e^-$ and $h^+$ accumulation in low-potential energy bands, redox capabilities diminish. In defective gap junctions, $e^-$ and $h^+$ remain in their SCs without charge exchange, therefore the two SCs operate independently without cooperation. A direct Z-scheme is produced when two SC materials with staggered band orientation come into close contact and charge is transferred among $e^-$ in CB of PS II and $h^+$ in VB of PS I (Fig. 4.3d). Despite sharing a staggered band alignment,

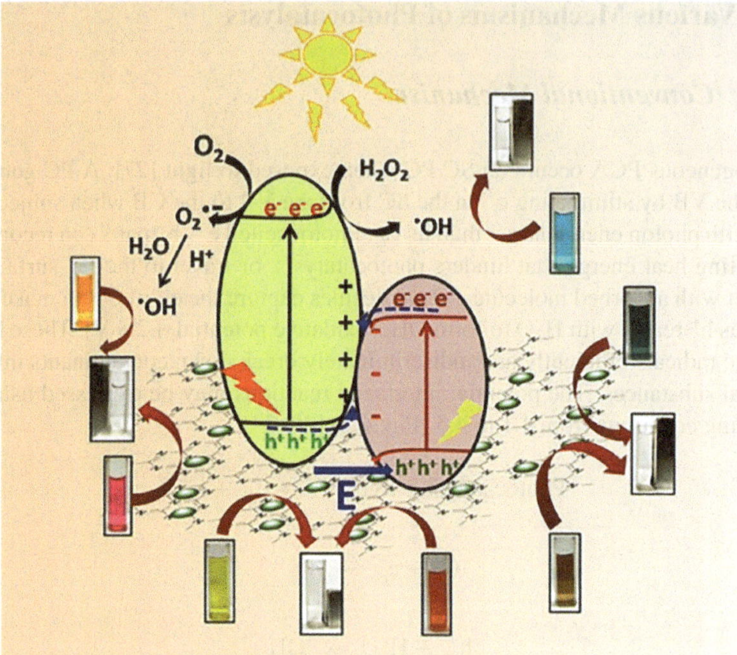


Fig. 4.2 Photocatalytic degradation of organic dyes. Reproduced with permission from [35], Copyright 2021, ACS

the Type II heterojunction and Z-scheme structures exhibit different paths of $e^-$ and $h^+$ migration. A Z-scheme heterojunction permits PS II photogenerated $e^-$ to shift and recombine with PS I photogenerated $h^+$, unlike Type II heterojunctions. The VB of PS II and the CB of PS I contain high-energy $h^+$ and $e^-$, which can be used for PCA. The $e^-$ and $h^+$ can participate in PCA reactions because this heterojunction type preserves their strong reductive and oxidative capacities. Therefore, direct Z-scheme PC generally exhibits better PCA compared to Type II heterogeneous PC [36].

#### 4.2.2.1 Straddling-Gap Junctions (Type I Mechanism)

In the type I process, the $e^-$–$h^+$ pairs created by light can chemically bond with dye molecules, resulting in the formation of excited dye (Dye*) that can be transformed into cationic ($Dye^{\bullet +}$) and anionic ($Dye^{\bullet -}$) radicals. This process ultimately leads to the production of reduction products. The oxidant can capture $e^-$ of CB from the excited dye after self-sensitized disintegration. This action restores the catalyst to its initial condition, so concluding the photocatalytic cycle. Modifying PC using dye sensitization works well. It expands the spectrum of light absorption to encompass visible

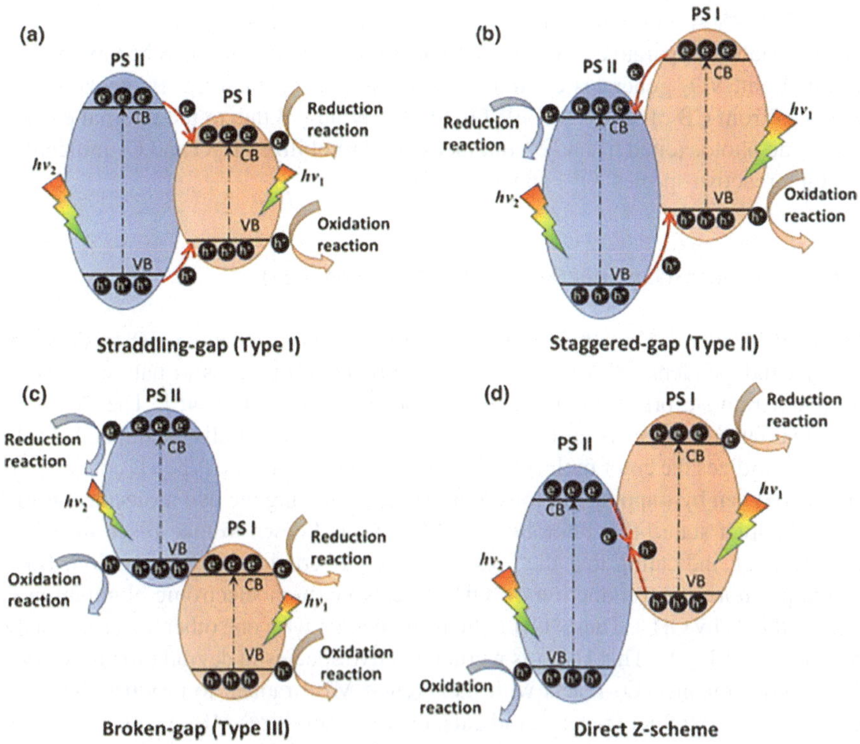


**Fig. 4.3** Charge flow directions in **a** Type I, **b** Type II, **c** Type III, and **d** direct Z-scheme heterojunction. Reproduced with permission from [36] Copyright 2021, Elsevier

light, enabling the extraction of solar energy. Additionally, it enhances the effectiveness of excitation and facilitates the separation of charges, resulting in improved PCA [37–39]. For instance, in the case of $ZnO/Fe_3O_4/g\text{-}C_3N_4$-50% nanocomposites, it is also proposed that there is a transfer of charge to the surface, which then immediately interacts with methyl orange (MO) dye [40].

#### 4.2.2.2 Staggered-Gap Junctions (Type II Mechanism)

In the type II process, a SC PC absorbs light photons, causing $e^-$ to be elevated from the VB to the CB, while simultaneously creating $h^+$ in the VB. The produced pairs of $h^+$ and $e^-$ are involved in reduction and oxidation processes. The $e^-$ and $h^+$ combine with oxygen and water ($H_2O$) to generate $O_2^{\cdot -}$ and $HO^{\cdot}$, respectively. These radicals then proceed to oxidize the dye molecules, resulting in the formation of $H_2O$ and $CO_2$ [37, 41]. The utilization of type II heterojunction PC is a highly successful method for improving the performance of PCA. This is achieved by efficiently separating the $e^-$–$h^+$ pairs formed during the process, through the band alignment between two SCs [42]. A $VO_2/CuWO_4$ NSs that is type II n–n heterojunction uses n-type SC materials to separate $e^-$–$h^+$ pairs and decrease recombination. When exposed to visible light, $VO_2$ generates $e^-$ and $h^+$ as a result of its small Eg. Photoexcited $e^-$ migrates from CB of $VO_2$'s to $CuWO_4$'s lower lying CB due to the internal electric field. The photoexcited $h^+$ and $e^-$ interact with immobilized $H_2O$ and $O_2$ molecules, forming reactive species $^{\cdot}O^{2-}$ and $^{\cdot}OH$ [43].

#### 4.2.2.3 Broken-Gap Junctions (Type III Mechanism)

Direct PCA of dyes (methylene blue, methyl orange, rhodamine B) is slow and catalyst independent. PCA is a common dye reduction process in nature, however it cannot remove organic contaminants from wastewater [44, 45]. The PCA and photooxidation of dyes occurred during photolysis. The initially stimulated singlet oxygen oxidizes the color molecule. The fundamental state of oxygen is converted to singlet oxygen by trapping photons with energy matching the dye molecule's initial excited triplet state [46]. Photolysis of MB dye is known to take place in the air environment, indicating that the presence of $O_2$ is crucial for deterioration. Basic media promotes the production of $^{\cdot}OH$ radicals via monoelectronic elimination of $MB^+$ radicals by $OH^-$. These $^{\cdot}OH$ radicals then react with one other, resulting in the production of $H_2O_2$. This $H_2O_2$ is a crucial active species in degradation processes. In a similar manner, $O_2$ reacts with the excited MB* radical to produce the $O_2^{\cdot-}$. Each of these extremely reactive radicals directly photolyzes MB dye [47]. RhB and other N dyes resist photolysis and breakdown reductively anaerobically, creating carcinogenic chemicals [48].

### *4.2.3 Z-Scheme Mechanism*

The direct Z-scheme heterojunction is distinguished by the directional movement of $e^-$ from the CB of semiconductor (SC) PS II to the VB of PS I, forming a route that resembles the letter "Z." In order for $e^-$ transmission in the Z-scheme to occur, the energy level at the interface among the PS I and PS II must be almost continuous, like that of a conductor. A Z-scheme heterojunction can form under many circumstances. The Z-scheme heterojunction is made by directly connecting PS I and PS II. This is achieved by rational realignment of the Fermi energy level, which occurs due to the redistribution of charges (Fig. 4.4a). This leads to the creation of an internal electric field that accelerates the recombination of photoexcited $h^+$ in the VB of PS I and photoexcited $h^+$ in the CB of PS II. Consequently, PS II may gather and sustain photoexcited $h^+$ in its VB, whereas PS I can accumulate and keep photoexcited $e^-$ in its CB [49–51]. The photogenerated $h^+$ in the VB of PS I and the photogenerated $e^-$ in the CB of PS II will transfer through the conductor and recombine, leading to a high concentration of $h^+$ in the VB of PS II and $e^-$ in the CB of PS I. This promotes enhanced redox reactions in SC that are in contact with a conductor, such as C or

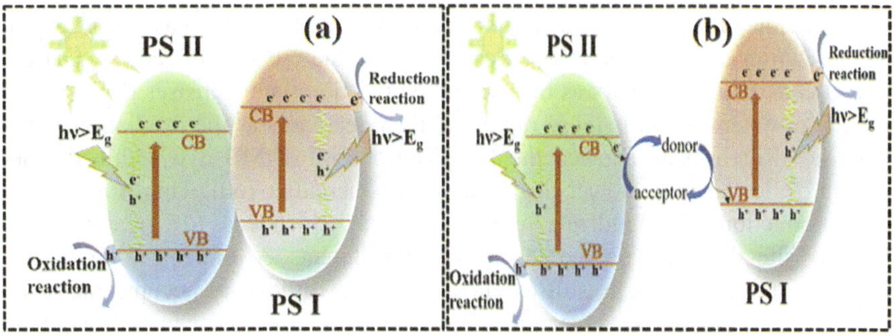


**Fig. 4.4** Diagram of (a) direct Z-scheme charge transfer and (b) redox-mediator Z-scheme charge transfer. Reproduced with permission from [57], Copyright 2019, Elsevier

metal-based materials. Furthermore, the $e^-$ acceptor and donor pair can operate as an $e^-$ mediator for the formation of Z-scheme junction between two SCs which are not directly connected to one other as shown in Fig. 4.4b [52].

Energy shortages and environmental contaminants can be addressed by direct Z-scheme photocatalytic devices. The light-converting ability and stability over time of each SC must be improved, as must the interaction surface between two components to increase PCA. A monocomponent SC PC has considerable challenges in overcoming the aforementioned obstacles in comparison to the artificial [53] Z-scheme photocatalytic device. An optimal choice for the reduction of contaminants, conversion of $CO_2$, and breakdown of $H_2O$ is a photocatalytic system that operates without the need for a redox mediator. This system follows a Z-scheme mechanism, which allows for minimal $e^-$ transmission distance. It works in liquid and gas phases. Redox-mediator Z-scheme photocatalytic devices have low photocatalytic effectiveness because of backward reactions, adverse reaction environments, ineffective long-distance e transfer, and light-shielding effects [54–56].

## 4.3 Photodegradation Rate-Affecting Parameters

Several parameters can affect the rate of PD, including irradiation reaction time, light intensity, medium composition, tap water quality, starting dye concentration, PC dose, oxidizing agent presence, and radical scavengers. The following points will be covered below.

### *4.3.1 Irradiation Time*

The duration of solar light exposure is a key element in influencing the photochemical capability of PC for dye degradation [58]. According to reports, a rise in the duration of irradiation time for photocatalytic reactions leads to an increase in PCA and decolorization of dye. This may be explained by the creation of radical $^{\cdot}OH$ and radical $O_2^-$, which intensify as the irradiation reaction time increases [59]. At first, the rate of deterioration is more rapid and then decreases as the irradiation period increases. Ramli et al. showed the fast deterioration of MB dye within the first 30 min, followed by a slower degradation rate thereafter [60]. This is due to the rapid production of $^{\bullet}OH$ radicals in the initial stage, which leads to a greater number of active sites available for the adsorption of pollutants. After a specific period of time, the active site that is still available becomes difficult to occupy due to the repulsive attraction between the molecules on the surface and those in the bulk phase. As a result, the rate of degradation tends to remain constant after a certain point [61]. Saeed et al. conducted an experiment where they utilized $TiO_2$/amidoxime-modified polyacrylonitrile nanofibers to degrade MB dye [62]. They observed that the degradation rate of the dye progressively rises with longer irradiation duration, reaching a plateau after 2 h.

### *4.3.2 UV Light Intensity*

The magnitude of light absorption by the SC catalyst at a specific wavelength is determined by the intensity of the light. The rate of initiation for PCA, namely the production of $e^-$–$h^+$ pairs in the photochemical process, is highly influenced by the intensity of light [63]. The spatial distribution of light intensity inside the reactor consistently governs the total conversion of pollutants and the efficiency of their degradation [64]. Many studies have examined the relationship between light intensity and the rate at which pollutants degrade. Some investigations have found that the degradation rate is proportional to the square root of the light intensity, while others have observed a linear relationship between the two variables [65]. Ollis et al. (1991) conducted a comprehensive analysis of the impact of light intensity on the pace at which organic pollutants break down [66]. According to Hermann (1999), the rate is directly proportional to the radiant flux $\Phi$ when $\Phi$ is less than 25 mW/cm$^2$ [67]. However, when the flux exceeds 25 mW/cm$^2$, the rate varies as $\Phi^{1/2}$, suggesting that the flux is too high and causing a rise in the $e^-$–$h^+$ recombination rate. The response rate remains constant regardless of the intensity of light at high levels. This is likely due to the prevalence of low-intensity reactions that involve the production of $e^-$–$h^+$ pairs, whereas the recombination of $e^-$–$h^+$ pairs is negligible. Kaneco et al. (2004) examined how variations in light intensity impact the process of solar PCA of Bisphenol A in water using $TiO_2$, specifically on both bright and overcast days [68]. The degrading efficiency demonstrates a quick increase when the light intensity rises up to 0.35 mW/cm$^2$, followed by a steady increase in efficiency.

Muruganandham et al. (2006) explored how UV light decolorizes and degrades Reactive Yellow 14. Decolorization and degradation rise between 16 and 32 W, but not significantly between 32 and 48 W, and it is independent of light intensity between 48 and 64 W [69]. The best radiation intensity was 30 W. In an 80-cm immobilized continuous flow reactor, increasing light intensity from 18.8 to58.5 $Wm^{-2}$ decolorizes Acid Red 27 from 25 to 42% and degrades it from 13 to 28% [70]. Venkatachalam et al. examined 4-CP mineralization efficiency employing 365 and 254 nm lights over $TiO_2$ [71]. Mineralization is somewhat greater at 365 than 254 nm. As light intensity increases from 5,000 to 45,000 lx, Acridine Orange degradation rate constant increases from 0.018 to 0.0892 $min^{-1}$ [69]. Chiou et al. examined phenol degradation under 20–400 W UV radiation. The degradation rate constants for UV/$TiO_2$ at 20, 100, and 400 W are $8.3 \times 10^{-3}$, 0.012, and 0.031 $min^{-1}$, respectively. An acceptable linear connection exists between light intensity and perceived first-order rate constant under the investigated conditions [72]. The PCA of Acid Red 88 utilizing immobilized ZnO increased with light intensity. At greater light intensity, the catalyst absorbs more photons, forming more $e^-{-}h^+$ pairs on the catalyst surface, increasing hydroxyl radical concentration and degradation rate [73].

### *4.3.3 Pollutant Type and Concentration*

To achieve effective implementation of a photocatalytic oxidation system, it is necessary to study the relationship between the rate of PCA and the concentration of the substrate. The study conducted by Selvam et al. shown that the degradation rate of 4-fluorophenol decreased as the starting concentration increased from 0.022 to 0.09 mM in the presence of $TiO_2$ and ZnO [74]. Phenomena of the same nature have been noticed in the process of breaking down phenol and m-nitrophenol through PCA, Acid Red 114, methyl red, and Acid Blue 80 when exposed to solar radiation, Amido Black, and Acid Orange 7. According to Tsai et al. (2009), the removal of bisphenol A reduced from 100 to 97% when the initial substrate concentration rose from 0.001 to 0.018 mM. Furthermore, the removal efficiency decreased from 97 to 67% when the BPA concentration was further increased to 0.044 mM [75–81].

Venkatachalam et al. found that 4-CP PCA rises up to 0.194 mM and subsequently declines from 0.194 to 0.233 mM. This impact is linked to screening effects at 4-CP concentrations over 0.194 mM [71]. PCA of Chromotrope 2B, Xylenol Orange, Bromothymol, Acridine Orange, Acid Orange 8, Amaranth, and Remazol Brilliant Blue R yields consistent findings [82]. Adsorbed reactant molecules may occupy all catalytic active sties at high pollutant concentration, slowing breakdown. According to various studies, more target pollutant molecules are adsorbed on the PC as concentration increases. Thus, the necessity for reactive species ($^{\cdot}OH$ and $^{\cdot}O^{-2}$) for pollutant degradation rises. For a given light intensity, catalyst quantity, and irradiation time, the production of $^{\cdot}OH$ and $^{\cdot}O^{-2}$ on the catalyst surface stays constant. Thus, OH radicals are insufficient for pollutant breakdown at greater concentrations. Subsequently, pollutant decomposition diminishes with concentration [83]. Increased substrate concentration can also produce intermediates that may adsorb on the catalyst. Slow diffusion of intermediates from the catalyst surface can deactivate PC active sites, slowing deterioration. According to apparent first-order kinetics, the rate of degradation is proportional to substrate concentration at low concentrations, regardless of catalytic site number [74]. Bismarck degradation increased with substrate concentration at the circumstances tested [84]. It follows the L–H law. The L–H kinetics model accurately represented PCA rates based on organic component concentration in several studies [85, 86]. Turchi and Ollis (1990) developed the L–H model to explain how starting solute concentrations affect response rates [87, 88].

### 4.3.4 *Effect of pH of the Medium*

pH is a significant constraint in PCA of dye reduction, since it has the potential to influence the reaction rates of the dye in several ways. The adsorption of dye on the surface of a SC is influenced by the pH of the solution, since it determines the charge of the catalyst surface [89]. A multitude of studies have been conducted in the past to investigate the impact of pH on the process of PCA of dyes [90–94] [95].

Investigating the impact of pH on the effectiveness of the PCA process is a complex undertaking, primarily because there are three potential reaction mechanisms that may lead to the reduction of the dye. These mechanisms include the attack of OH, oxidation by $h^+$, and reduction by $e^-$ in the CB. The influence of each process is contingent upon the pH and characteristics of the substrate [96]. Zhu et al. conducted a study on the relationship between pH and the surface charge characteristics of PC. They found that this relationship may be explained by the concept of point of zero charge (PZC) [97]. The $pH_{pzc}$ value for $TiO_2$ particles is 6.8 [98]. The PC surface acquires a positive charge in an acidic solution or when the pH is below a certain threshold (pH < 6.8), and vice versa [99], as indicated by the following equations:

$$TiOH_{(surface)} + H^+ \rightarrow TiOH_{2(surface)}$$

$$+ TiOH_{(surface)} + OH^- \rightarrow TiO^-_{(surface)} + H_2O$$

Positively charged adsorbent surfaces repel/attract cations when pH decreases below the PZC. The surface repels anion and attracts cation, unlike PZC. Bahnemann et al. observed metal oxide surface acid–base affects PCA [100].

The pH level has a significant impact on both the overall PCA process and the rate at which decolorization occurs. For example, the $h^+$ generated at lower pH levels functions as significant oxidation agents. $OH^\cdot$ are responsible for the oxidation process when the pH is higher or neutral [101]. They found that the pH has a negligible impact on the rate at which these substances vanish. At elevated pH levels, the process of adsorption of MB is favored over the interaction with a surface that has a negative charge. Conversely, the presence of a negatively charged sulfonate group in OG leads to a decrease in adsorption at high pH levels. Additional dyes, which displayed diverse functional groups, were shown to demonstrate a comparable to MB [102].

Baran et al. conducted a research on the degradation of Bromocresol purple dye and found that it degraded more effectively under acidic settings than to alkaline conditions [94]. An adsorption effectiveness increase of six times was observed following the acidification of the solution from pH 8.0 to 4.5. The increase in adsorption effectiveness may have been due to a pH alteration, rather than only changes in the $TiO_2$ surface [103]. Wei and Wan conducted a study on the impact of catalyst concentration on the rate of PCA, finding that it can have both beneficial and detrimental effects. The measured early reaction rates exhibited a direct proportionality to the quantity of catalyst, indicating the presence of a heterogeneous regime. To examine the degradation rates, several quantities of Degussa P25 ranging from 0.5 to 5 g/l were utilized by raising the pH of the solution above 10. It was discovered that a significant portion of the dye molecules that were adsorbed were not directly in touch with the PC surface. As a result, a drop in the rate of decolorization was seen as the pH fell [104]. Hu et al. studied pH-dependent reduction and adsorption of Cationic Blue X-GRL (CBX) and Procion Red MX-5B (MX-5B). They evaluated MX-5B adsorption reduction with pH. However, UV source PCA of MX-5B increased with decreased adsorption. The results revealed that MX-5B degraded first by hydroxylating more active C–N bonds from the naphthalene or benzene rings. Thus, MX-5B PCA in bulk aqueous phase. However, pH promoted CBX adsorption and degradation [105]. In acidic or basic circumstances, Orange G deteriorated faster than in neutral settings, according to Hung and Yuan. Higher decolorization under basic circumstances may result from increased $OH^-$ in the solution, leading to more $OH^\cdot$ radicals [106].

### 4.3.5 *Effect of Reaction Temperature*

Increasing temperature slowly improves organic compound PCA. Free radicals are produced by solution bubbles as temperature rises. Temperature increases accelerate deterioration by exceeding $e^-$–$h^+$ recombination. By raising the temperature, organic molecules oxidize faster at the interface [107].

Mozia et al. conducted a study on decolorization by Acid Red 18 and discovered that PCA increases as the reaction temperature is raised [108]. Soares et al. also documented that optimal working temperature of around 40–50 °C for the reaction. Due to its slower rate than surface degradation and reactant adsorption, product desorption slows the reaction at low temperatures. The essential stage is dye adsorption on the PC, which happens when the temperature is raised [109]. Zhou and Ray proposed that the drop in the rate constant is caused by a loss in the adsorptive capacity associated with $O_2$ and organic substances at higher temperatures [110]. Therefore, the ideal temperature often falls within the range of 293 to 353 K [111].

## 4.4 Methods for Improving the Photocatalytic Efficacy

Numerous variables can greatly impact a photocatalyst's PCA. The crystal's composition and physical structure are among these factors. Ideal features include visible light capture, high charge separation efficiency, and minimal charge recombination. It should also have a low Eg for easy photon activation. These traits make it a promising PC. High Eg in minerals like TiO2 require UV photons for stimulation, as visible light is insufficient. In addition, these materials exhibit significant recombination rates due to inefficient removal of photoexcited $e^{-}$–$h^{+}$ pairs. In recent years, doping, heterostructure, and shape manipulation have been developed to circumvent these limitations.

### *4.4.1 Metal Doping*

Second-generation visible-light-active PC ($TiO_2$ and ZnO) have been doped with metals to enhance their PCA in visible light. This was accomplished by adding a metal ion to the SC crystalline structure. The incorporation of various metal ions and precious metals (Ag, Au, and Pt) into the $TiO_2$ PC is a highly successful approach for enhancing its catalytic activity [112–114]. Adding metal ions to $TiO_2$ changes its energy level structure. Metals' increased reactivity makes $e^{-}$s excitation easier.

As a result, the absorption range in a $TiO_2$ system becomes broader [114, 115]. Figure 4.5 demonstrates that metals have the ability to catch the $e^{-}$ produced by the excitation of $TiO_2$. Furthermore, the $e^{-}$ within $TiO_2$ have a limited tendency to revert back to their original form. Following the introduction of metal ions, metal doping, can assist as centers for trapping carriers [116, 117]. Metals with many valence states can have d orbitals' $e^{-}$ transition and penetrate the $TiO_2$ lattice. This procedure results in a decrease in the Eg and the energy required for $e^{-}$ movement [118–120].

In 1994, Choi et al. initiated research on the integration of metal ions into $TiO_2$. The experiment involved doping 19 different metal ions into nano-$TiO_2$. The catalytic activity was shown to be improved when $Fe^{3+}$, $Mo^{5+}$, and $Rh^{3+}$ were used as dopants [121]. This improvement can be attributed to the matching of the ionic potential and radius of the doped ions with $TiO_2$. Du et al. utilized $Ge^{4+-}$ incorporated $TiO_2$ to fabricate Ge/$TiO_2$ PC and successfully decomposed ciprofloxacin. The $Ge^{4+}$ ion had a radius of 0.054 nm, which was less than the radius of the $Ti^{4+}$ ion (0.068 nm). The introduction of Ge into the $TiO_2$ lattice resulted in the substitution of Ti atoms, leading to lattice defects and a subsequent delay in the recombination of $e^{-}$ and $h^{+}$ [122]. Consequently, the generation of $^{\bullet}OH$ on the $TiO_2$ surface led to an enhancement in PCA. At 0.26% doping and 571 °C calcination, the PCA was 97%. Crisan et al. synthesized iron-$TiO_2$ by the sol–gel technique, resulting in an absorption spectrum that reached a wavelength of 546 nm. Furthermore, the band width at the specific wavelength of 410 nm was measured to be 3.03 eV when the iron concentration (w)

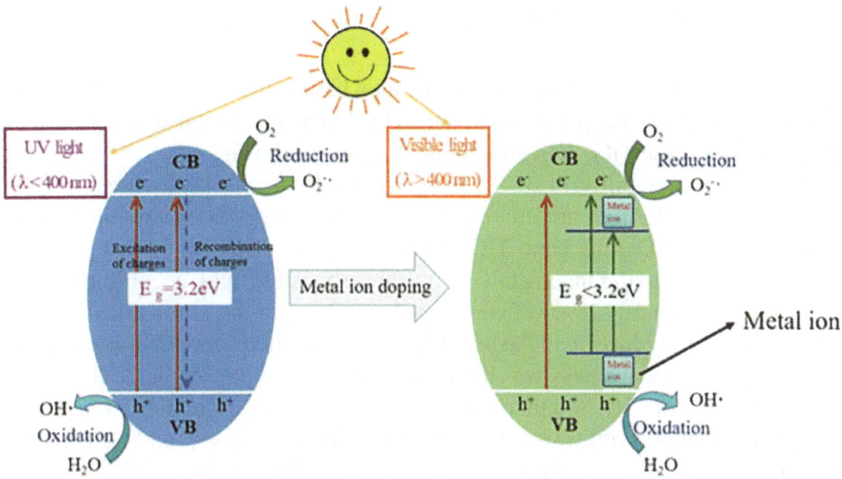


**Fig. 4.5** Mechanism of metal ion-doped $TiO_2$. Reproduced with permission from [126], Copyright 2020, catalysts

was 2%. The addition of 0.5% Fe nano-$TiO_2$ raised the breakdown rate of nitrobenzene from 70 to 88% compared to pure $TiO_2$ [123]. Gnanasekaran et al. found that decreasing the Eg width to 3.01 eV extended absorption range of Co-$TiO_2$'s spectrum from 382 to 411 nm. The photocatalytic treatment using visible light resulted in a degradation rate of 53% for MO after 240 min, which enhanced its PCA [124]. Huang et al. synthesized Mo-$TiO_2$ powders using sol–gel. When exposed to natural sunlight, the degradation efficiency of MB was found to be 98%, using a $Mo^{6+}$ concentration of 2% (w). The primary reason for this outcome was the decrease in the Eg from 3.05 to 2.73 eV as a result of $Mo^{6+}$ doping, together with the broader range of wavelengths absorbed during excitation [125]. Bhatia and Dhir synthesized Ni- and Bi-doped $TiO_2$ by the sol–gel technique and observed that Bi-$TiO_2$ achieved a maximum ibuprofen degradation rate of 89%, while Ni-$TiO_2$ reached 78%. The observed result may be ascribed to the augmentation of the particular surface area and the reduction in the size of the crystallite. Wang et al. synthesized a catalyst consisting of $TiO_2$ nanotube arrays doped with $Fe^{3+}$ using a straightforward hydrothermal technique. This resulted in a 20% enhancement in the breakdown rate of MB. The study further discovered that the introduction of $Fe^{3+}$ into the $TiO_2$ lattice created traps, resulting in a significant enhancement of the separation efficacy of $e^-$–$h^+$ pairs [126].

### 4.4.2 Non-metal Doping

A prominent topic in photocatalytic modification has always been non-metal doping. $TiO_2$'s photoresponse range can be expanded by doping non-metal atoms. Non-metal ion doping raises the $TiO_2$ VB, shortening the gap width (Fig. 4.6).

The long-term stability of $e^-$–$h^+$ pairs created by light and the area of $TiO_2$'s sensitivity to visible light are crucial factors in enhancing its efficacy in pollution remediation [127]. Doping nano-$TiO_2$ with non-metals including S, C, I, and F has enlarged its wavelength range, allowing it to absorb visible light, according to recent study. The non-metal [128–130] was doped, resulting in the reconstruction of the energy band and a reduction in the breadth of the prohibited band. Consequently, the range of light responsiveness was increased [131–133]. First, Asahi et al. replaced a minor fraction of $TiO_2$'s lattice oxygen with $N_2$ doping to achieve visible light PCA [134]. Li et al. synthesized N-doped $TiO_2$ through sol–gel, and observed that pure $TiO_2$ showed 5% reduction whereas N-doped $TiO_2$ revealed 95% reduction at 180 min [135]. Because N impurity and $Ti^{3+}$ narrowed Eg of $TiO_2$'s, N doping greatly improved MO degradation under visible light. Jyothi et al. hydrothermally synthesized N and F co-doped $TiO_2$ to eliminate bromoethane [136]. The synergistic impact boosted bromoethane elimination from 54 and 94% for pure and doped $TiO_2$. Rahbar et al. hydrothermally produced S- and N-doped CQDs/$TiO_2$ composites. Acidic AR88 (azo dyes) degraded 54% faster than pure $TiO_2$ under visible light. They also found that $e^-$ transport in CQDs separated charges. The surface

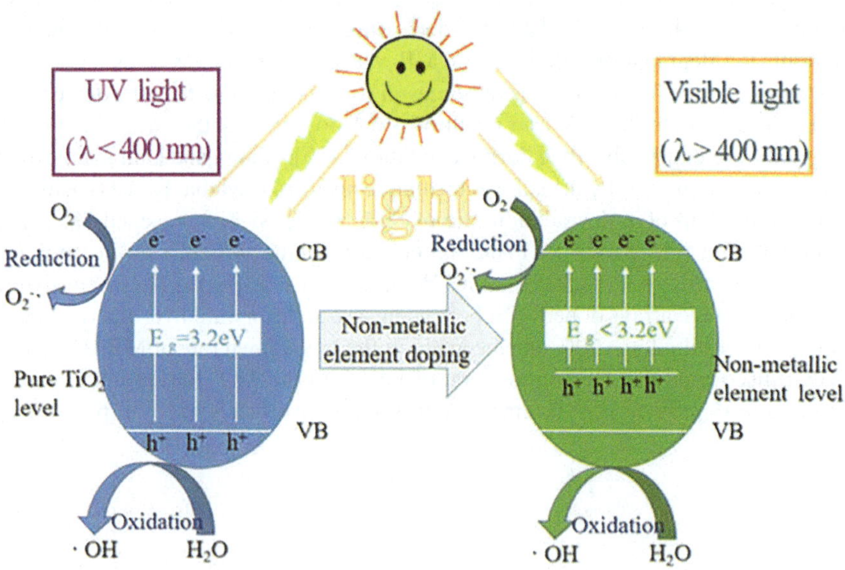


**Fig. 4.6** The mechanism of non-metallic element doping modification system. Reproduced with permission from [126], Copyright 2020, catalysts

functional group also increased pollutant molecule adsorption on the PC surface, promoting PCA [137]. Liu et al. employed I-$TiO_2$ as a photoelectrocatalytic electrode. After 2 h (1.4 V) of visible light, the improved photoelectrode eliminated 60% of diclofenac, whereas the $TiO_2$ photoelectrode removed 10%. CNTs' high electron storage capacity allows them to receive photoexcited $e^-$ in $TiO_2$'s CB. Therefore, doping non-metal elements expands the photocatalytic process to visible light and enhances $TiO_2$ photon efficiency by separating charge carriers [138].

Doping non-metallic materials increases $TiO_2$'s visible light sensitivity but narrows its Eg. Therefore, the $TiO_2$ nanocrystalline phase's oxidizing activity is immediately diminished, and adsorbed compounds cannot be totally destroyed. Non-metal doping in PCA is a hot and demanding topic.

## 4.5 Carbon-Based Photocatalysts

### *4.5.1 0D CBMS Photocatalysts*

The newly discovered C allotropes, namely, CQDs, CDs, and GQDs, are C NPs that have a quasi-spherical shape and are uniformly sized, with diameters smaller than 10 nm. CQDs exhibit direct absorption of light within the UV–vis region due to the transition of $e^-$ from HOMO to the LUMO facilitated by the presence of $sp^2$ C=C bonds. CQDs absorb mostly in the visible spectrum due to $n-\pi^*$ transitions caused by the existence of functional groups. CQDs exhibit distinct characteristics due to the phenomenon of quantum confinement, which causes their Eg to vary with their sizes. Due to their distinct structural characteristics, CQDs have remarkable abilities to capture sunlight, adjustable PL properties, and efficient transmission of photoexcited $e^-$. These qualities make them suitable for use in PCA [139]. Furthermore, the photoexcitation of CDs is a result of the absorption of $\pi$-plasmons in the core C NPs [140, 141].

The utilization of natural resources for the preparation of CQDs offers the benefits of cost-effectiveness and environmental sustainability. It is urged to promote the creation of innovative approaches that utilize natural C sources. In addition, GQDs doped with heteroatoms such as sulfur and $N_2$ were physically mixed with $TiO_2$ to form composites known as S- and N-QGD/$TiO_2$. These composites were used for the photocatalytic degradation of RhB [142]. Furthermore, it has been shown that both ZnO/S and N-GQDs demonstrate remarkable efficacy in degrading dyes, as seen by their high dye degradation efficiencies [80]. CQDs serve many roles in efficient PCA: (i) acting as a acceptor, mediator, and for $e^-$ generated in the CB; (ii) enhancing the PC visible light activity; (iii) functioning as a reductant in the development of surface plasmon resonance process of various metal NPs; and (iv) efficiently capturing a wide range of solar wavelengths by utilizing PL. Hence, the NPs, VB, and CB are interdependent in determining the excitation and PCA of PC that have been changed with CQDs [143]. Due to their high crystallinity, GQDs exhibit enhanced $e^-$ transport characteristics and extended carrier lifetimes. Furthermore, $C_{60}$ is widely regarded as a highly proficient $e^-$ acceptor and transporter, possessing a low Eg of around 1.6–1.9 eV. They can be utilized as direct energy transfer mediators to generate singlet oxygen for the process. Due to their distinctive composition, they possess significant use in several domains, including photovoltaics and PCA [144, 145].

### *4.5.2 1D CBMs Photocatalysts*

CNTs are excellent choices for supporting NMs due to their combined properties of nanoscale dimensions and fibrous structure. Due to their extensive surface areas, fascinating electrical characteristics, distinctive physicochemical qualities, and elevated aspect ratios, CNTs show great potential as suitable options for developing and creating innovative PC. Typically, in PC systems based on CNTs, the CNTs serve as "electron sinks" that collect $e^-$ produced by the SC. This allows for the efficient separation of $e^-$–$h^+$ pairs [146]. Additionally, they enable $e^-$ transit in heterogeneous PC systems via the use of "electron bridges." Notably, several researchers have noticed that the absorption edge of PC based on CNTs changes toward longer wavelengths compared to PC. CNTs can also undergo functionalization through processes such as surface oxidation or doping of atom. This greatly simplifies the strong bond and uniform dispersion of SC NPs on the CNTs. SWCNTs are 1D quantum wires with good conductivity with tube diameter is smaller than 6 nm [147]. It is important to mention that untreated CNTs consist of several forms of disordered C, shards of Gt, and other contaminants, which hinders their suitability for extensive use. Furthermore, it poses a challenge to integrate them with other SC in order to synthesize heterostructures. The crucial aspect in the context of CNTs is the interfacial connection between the CNTs and another SC. This connection promotes the flow of $e^-$ and diminishes the recombination of $e^-$–$h^+$ pairs. Hence, it ought to undergo surface oxidation or be subjected to metal/non-metal doping in order to enhance its functionality. So far, researchers have developed many PC based on CNTs that can harness solar energy to break down different contaminants into harmless compounds [148].

### *4.5.3 2D CBNMs Photocatalysts*

Due to its unique properties, graphdiyne (GDY) is used in PCA showed high performance [149]. The range of ligands gives graphdiyne structural differences and customizable properties. These semiconducting materials have Eg of 0.46–1.22 eV. Their vast surface areas and porous structure accelerate mass transfer and expose more catalytic sites, boosting catalytic reactions. Wong et al. reported that these materials are chemically stable throughout catalytic reactions in acidic and alkaline environments [150]. Thus, graphdiyne-based compounds are photocatalytic platforms. Many experimental researches have developed GDY-based heterojunctions with other SCs, such as $TiO_2$, to remove photocatalytic organic pollutants. Graphdiyne and $TiO_2$ NPs created a Ti–O–C connection. The composite's PCA for MB decolorization was induced by carbon p orbital impurity bands. Graphdiyne accepted photogenerated $e^-$s, reducing charge recombination and speeding dye breakdown [151].

In 2006, g-$C_3N_4$ was developed as a PC in heterogeneous catalysis because of its excellent chemical and thermal stability, Eg of 2.7 eV, CB of −1.1 eV, and VB of 1.6 eV positions compared to hydrogen electrode. In PCA, SC creates $e^-$–$h^+$ pairs with light, requiring active substances like $H^+$, $^{\cdot}OH$, and $^-$, $^{\cdot}O_2^-$ radicals are formed when $O_2$ combines with $e^-$. Two mechanisms yield $^{\cdot}OH$ radicals: photoexcited $h^+$ directly oxidizing $H_2O/OH \times$ (1.99 eV oxidation potential) and dispersion reaction of $^{\cdot}O_2^-$ radicals. Its smaller surface area, elevated recombination rate, and lower quantum yield owing to weak van der Waals connections among g-$C_3N_4$ layers limit its practical use [148]. Thus, researchers have altered g-$C_3N_4$ coupling with other SC to vary its electrical properties by band assembly engineering. Increasing g-$C_3N_4$ crystallinity reduces photogenerated charge recombination and improves PCA [152]. The PCA is highly reliant on shape and size, which can increase absorbance of visible light [153]. Many ways have been explored to increase g-$C_3N_4$'s PCA. g-$C_3N_4$ has been thermally polymerized from nitrogen-rich organic precursors by several authors. Material photoactivities are significantly correlated with microstructure. The catalyst microstructure and catalyst-associated reaction pathway synergetic reactions were elucidated. Thus, nanocomposites' components interact depending on their production techniques and circumstances. Surface alteration and defects in g-$C_3N_4$ enhanced light absorbance and prevented charge carrier recombination. Breakage and interruption of g-$C_3N_4$'s fundamental unit configuration affects PCA. A potential visible light PC for water remediation is N-defective g-$C_3N_4$ with an adjustable band structure. Structures, such as nanosheet, nanowire, nanotube, nanorod, and nanobelt shapes, diameters, and sizes, affect g-$C_3N_4$ activity. 2D sheets, well-defined building units, are more important in PCA than bulk g-$C_3N_4$. Second, doping g-$C_3N_4$ with metal/non-metal components modifies light absorption capacity. The third technique for PCA involves heterojunctions formed by g-$C_3N_4$ and other SC PC with acceptable energy levels [148].

The incorporation of CDs into g-$C_3N_4$ notably increased its PCA. Due to their tiny size, capacity to absorb visible light, and ability to convert light into higher energy states, CDs possess $e^-$ transport and storage capabilities, making them highly suitable for PCA when paired with another SC material. In recent times, several groups have developed metal-free CDs/g-$C_3N_4$ hybrid PC for diverse PCA [154, 155]. Li et al. prepared CQD-decorated g-$C_3N_4$ using an innovative approach to degrade RhB under visible light exposure [156]. The analysis demonstrated that the rate at which the dye decolorize in the g-$C_3N_4$/CQDs nanocomposites was 0.013 $moL^{-1}$ $min^{-1}$, which is three times higher than the rate seen in the bare g-$C_3N_4$ (0.0044 $moL^{-1}$ $min^{-1}$),

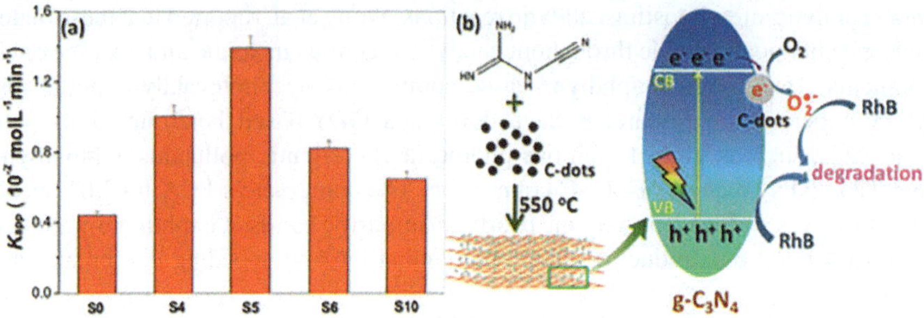


**Fig. 4.7** **a** Comparison of the degradation rate constant of C-dots-modified g-$C_3N_4$ **b** diagram showing the one-pot synthesis of C-dots-modified g-$C_3N_4$ with enhanced PCA. Reproduced with permission from [148]. Copyright 2023, Molecules

as illustrated in Fig. 4.7a, b. The improved PCA can be attributed to the assistance of CDs in facilitating the effective transport of $e^-$ from the CB of g-$C_3N_4$ to the adsorbed $O_2$, resulting in the formation of $O_2^{\cdot}$ [148].

## References

1. Nagajyothi, P.C., et al.: Green synthesis: photocatalytic degradation of textile dyes using metal and metal oxide nanoparticles-latest trends and advancements. Crit. Rev. Environ. Sci. Technol. **50**(24), 2617–2723 (2020)
2. Cao, Z., et al.: Doping of chlorine from a neoprene adhesive enhances degradation efficiency of dyes by structured $TiO_2$-coated photocatalytic fabrics. Catalysts **10**(1), 69 (2020)
3. Zheng, Y., et al.: Microscale flower-like magnesium oxide for highly efficient photocatalytic degradation of organic dyes in aqueous solution. RSC Adv. **9**(13), 7338–7348 (2019)
4. Vallejo, W., Cantillo, A., Díaz-Uribe, C.: Methylene blue photodegradation under visible irradiation on Ag-Doped ZnO thin films. Int. J. Photoenergy **2020** (2020)
5. Melinte, V., Stroea, L., Chibac-Scutaru, A.L.: Polymer nanocomposites for photocatalytic applications. Catalysts **9**(12), 986 (2019)
6. Ola, O., Maroto-Valer, M.M.: Review of material design and reactor engineering on $TiO_2$ photocatalysis for $CO_2$ reduction. J. Photochem. Photobiol., C **24**, 16–42 (2015)
7. Pouretedal, H., Kiyani, M.: Photodegradation of 2-nitrophenol catalyzed by CoO, CoS and CoO/CoS nanoparticles. J. Iran. Chem. Soc. **11**, 271–277 (2014)
8. Saeed, K., et al.: Synthesis, characterization and photocatalytic activity of silver nanoparticles/amidoxime-modified polyacrylonitrile nanofibers. Fibers Polym. **16**, 1870–1875 (2015)
9. Ahmed, S., et al.: Advances in heterogeneous photocatalytic degradation of phenols and dyes in wastewater: a review. Water Air Soil Pollut. **215**, 3–29 (2011)
10. Ray, M.B.: Photodegradation of the volatile organic compounds in the gas phase: a review. Dev. Chem. Eng. Miner. Process. **8**(5–6), 405–439 (2000)
11. Saeed, K., et al.: $TiO_2$/carbon nanotubes composites used for the photocatalytic degradation of organic dye. J. Chem. Eng. Chem. Res. **2**, 671–676 (2015)
12. Julson, A.J., Ollis, D.F.: Kinetics of dye decolorization in an air–solid system. Appl. Catal. B **65**(3–4), 315–325 (2006)
13. Zada, N., Saeed, K., Khan, I.: Decolorization of Rhodamine B dye by using multiwalled carbon nanotubes/Co–Ti oxides nanocomposite and Co–Ti oxides as photocatalysts. Appl. Water Sci. **10**, 1–10 (2020)

14. Pornrungroj, C., Onodera, T., Oikawa, H.: PCBM nanoparticles as visible-light-driven photocatalysts for photocatalytic decomposition of organic dyes. MRS Commun. **9**(1), 321–326 (2019)
15. Song, S., et al.: Synthesis of Z-scheme multi-shelled ZnO/$AgVO_3$ spheres as photocatalysts for the degradation of ciprofloxacin and reduction of chromium (VI). J. Mater. Sci. **55**, 4987–5007 (2020)
16. Garrido, I., et al.: Photocatalytic performance of electrospun silk fibroin/ZnO mats to remove pesticide residues from water under natural sunlight. Catalysts **10**(1), 110 (2020)
17. Tamimi, M., et al.: Photocatalytic degradation of pesticide methomyl: determination of the reaction pathway and identification of intermediate products. Photochem. Photobiol. Sci. **5**, 477–482 (2006)
18. Marquès, M., et al.: Photodegradation of polycyclic aromatic hydrocarbons in soils under a climate change base scenario. Chemosphere **148**, 495–503 (2016)
19. Wang, Y., et al.: Photodegradation of polycyclic aromatic hydrocarbon pyrene by iron oxide in solid phase. J. Hazard. Mater. **162**(2–3), 716–723 (2009)
20. Liu, B., et al.: Photocatalytic degradation of polycyclic aromatic hydrocarbons in offshore produced water: effects of water matrix. J. Environ. Eng. **142**(11), 04016054 (2016)
21. Li, J., et al.: Photodegradation of benzothiazole ionic liquids catalyzed by titanium dioxide and silver-loaded titanium dioxide. Chin. J. Chem. Eng. **28**(5), 1397–1404 (2020)
22. Gu, J., et al.: Study on degradation of benzothiazolium-based ionic liquids by UV-$H_2O_2$. Appl. Sci. **10**(3), 894 (2020)
23. Zeng, X., et al.: Photodegradation of polyfluorinated dibenzo-p-dioxins in organic solvents: experimental and theoretical studies. Environ. Sci. Technol. **50**(15), 8128–8134 (2016)
24. Li, Y., et al.: Photocatalytic water splitting by N-$TiO_2$ on MgO (111) with exceptional quantum efficiencies at elevated temperatures. Nat. Commun. **10**(1), 4421 (2019)
25. Qamaruddin, M., et al.: Sulfur doped ceria-titania (S-$CeTiO_{4-x}$) nanocomposites for enhanced solar-driven water splitting. Sol. Energy **188**, 890–897 (2019)
26. Khan, I., et al.: Heterogeneous photodegradation of industrial dyes: an insight to different mechanisms and rate affecting parameters. J. Environ. Chem. Eng. **8**(5), 104364 (2020)
27. Xu, C., et al.: Nanostructured materials for photocatalysis. Chem. Soc. Rev. **48**(14), 3868–3902 (2019)
28. Saeed, K., Khan, I., Sadiq, M.: Synthesis of graphene-supported bimetallic nanoparticles for the sunlight photodegradation of Basic Green 5 dye in aqueous medium. Sep. Sci. Technol. **51**(8), 1421–1426 (2016)
29. Chiu, Y.-H., et al.: Mechanistic insights into photodegradation of organic dyes using heterostructure photocatalysts. Catalysts **9**(5), 430 (2019)
30. Anwer, H., et al.: Photocatalysts for degradation of dyes in industrial effluents: opportunities and challenges. Nano Res. **12**, 955–972 (2019)
31. Mandal, B., et al.: $MnFe_2O_4$ decorated reduced graphene oxide heterostructures: nanophotocatalyst for methylene blue dye degradation. Vacuum **173**, 109150 (2020)
32. Al Kausor, M., Gupta, S.S., Chakrabortty, D.: Ag3$PO_4$-based nanocomposites and their applications in photodegradation of toxic organic dye contaminated wastewater: review on material design to performance enhancement. J. Saudi Chem. Soc. **24**(1), 20–41 (2020)
33. Tentu, R.D., Basu, S.: Photocatalytic water splitting for hydrogen production. Curr. Opin. Electrochem. **5**(1), 56–62 (2017)
34. Mortazavian, S., Saber, A., James, D.E.: Optimization of photocatalytic degradation of acid blue 113 and acid red 88 textile dyes in a UV-C/$TiO_2$ suspension system: application of response surface methodology (RSM). Catalysts **9**(4), 360 (2019)
35. Gogoi, D., Makkar, P., Ghosh, N.N.: Solar light-irradiated photocatalytic degradation of model dyes and industrial dyes by a magnetic $CoFe_2O_4$–$gC_3N_4$ S-scheme heterojunction photocatalyst. ACS Omega **6**(7), 4831–4841 (2021)
36. Li, X., et al.: A review of material aspects in developing direct Z-scheme photocatalysts. Mater. Today **47**, 75–107 (2021)

37. Isac, L., et al.: Copper sulfide based heterojunctions as photocatalysts for dyes photodegradation. Front. Chem. **7**, 694 (2019)
38. Natarajan, S., Bajaj, H.C., Tayade, R.J.: Recent advances based on the synergetic effect of adsorption for removal of dyes from waste water using photocatalytic process. J. Environ. Sci. **65**, 201–222 (2018)
39. Zhao, J., Chen, C., Ma, W.: Photocatalytic degradation of organic pollutants under visible light irradiation. Top. Catal. **35**, 269–278 (2005)
40. Wu, Z., et al.: A ternary magnetic recyclable $ZnO/Fe_3O_4/gC_3N_4$ composite photocatalyst for efficient photodegradation of monoazo dye. Nanoscale Res. Lett. **14**, 1–14 (2019)
41. Al-Mamun, M., et al.: Photocatalytic activity improvement and application of UV-$TiO_2$ photocatalysis in textile wastewater treatment: a review. J. Environ. Chem. Eng. **7**(5), 103248 (2019)
42. Low, J., et al.: A review of direct Z-scheme photocatalysts. Small Methods **1**(5), 1700080 (2017)
43. Dashtian, K., et al.: Achieving enhanced blue-light-driven photocatalysis using nanosword-like $VO_2/CuWO_4$ type II n–n heterojunction. Chem. Eng. J. **339**, 189–203 (2018)
44. Zhang, G., Zhang, S.: Quantitative structure-activity relationship in the photodegradation of azo dyes. J. Environ. Sci. **90**, 41–50 (2020)
45. Pingmuang, K., et al.: Composite photocatalysts containing $BiVO_4$ for degradation of cationic dyes. Sci. Rep. **7**(1), 8929 (2017)
46. Xie, Y., Yuan, C.: Visible-light responsive cerium ion modified titania sol and nanocrystallites for X-3B dye photodegradation. Appl. Catal. B **46**(2), 251–259 (2003)
47. Soltani, T., Entezari, M.H.: Photolysis and photocatalysis of methylene blue by ferrite bismuth nanoparticles under sunlight irradiation. J. Mol. Catal. A: Chem. **377**, 197–203 (2013)
48. Rajagopalan, V.: A new synergetic nanocomposite for dye degradation in dark and light. Sci. Rep. **6**(1), 1–10 (2016)
49. Pan, T., et al.: Anionic polyacrylamide-assisted construction of thin 2D–2D $WO_3/g-C_3N_4$ Step-scheme heterojunction for enhanced tetracycline degradation under visible light irradiation. J. Hazard. Mater. **393**, 122366 (2020)
50. Jiang, Z., et al.: A hierarchical Z-scheme $\alpha-Fe_2O_3/g-C_3N_4$ hybrid for enhanced photocatalytic $CO_2$ reduction. Adv. Mater. **30**(10), 1706108 (2018)
51. Chao, Y., et al.: Ultrathin visible-light-driven Mo incorporating $In_2O_3$–$ZnIn_2Se_4$ Z-scheme nanosheet photocatalysts. Adv. Mater. **31**(5), 1807226 (2019)
52. Maeda, K.: Z-scheme water splitting using two different semiconductor photocatalysts. ACS Catal. **3**(7), 1486–1503 (2013)
53. Nayak, S., Parida, K.: Deciphering Z-scheme charge transfer dynamics in heterostructure NiFe-LDH/N-rGO/g-$C_3N_4$ nanocomposite for photocatalytic pollutant removal and water splitting reactions. Sci. Rep. **9**(1), 2458 (2019)
54. You, L., et al.: Investigation of the kinetics and mechanism of Z-scheme Ag 3 $PO_4/WO_3$ p–n junction photocatalysts with enhanced removal efficiency for RhB. New J. Chem. **43**(43), 17104–17115 (2019)
55. Yan, Y., et al.: Direct Z-scheme $CaTiO_3$@ BiOBr composite photocatalysts with enhanced photodegradation of dyes. Environ. Sci. Pollut. Res. **26**, 29020–29031 (2019)
56. Ma, Y., et al.: Enhanced visible-light photocatalytic degradation by $Mn_3O_4/CeO_2$ heterojunction: a Z-scheme system photocatalyst. Inorg. Chem. Front. **5**(10), 2579–2586 (2018)
57. Huang, D., et al.: Artificial Z-scheme photocatalytic system: what have been done and where to go? Coord. Chem. Rev. **385**, 44–80 (2019)
58. Makeswari, M., Saraswathi, P.: Photo catalytic degradation of methylene blue and methyl orange from aqueous solution using solar light onto chitosan bi-metal oxide composite. SN Appl. Sci. **2**(3), 336 (2020)
59. Sakthivel, S., et al.: Solar photocatalytic degradation of azo dye: comparison of photocatalytic efficiency of ZnO and $TiO_2$. Sol. Energy Mater. Sol. Cells **77**(1), 65–82 (2003)
60. Che Ramli, Z.A., et al.: Photocatalytic degradation of methylene blue under UV light irradiation on prepared carbonaceous. Sci. World J. **2014** (2014)

61. Aziztyana, A.P., et al.: Optimisation of methyl orange photodegradation using $TiO_2$-zeolite photocatalyst and $H_2O_2$ in acid condition. In: IOP Conference Series: Materials Science and Engineering. IOP Publishing (2019)
62. Saeed, K., Khan, I., Park, S.-Y.: $TiO_2$/amidoxime-modified polyacrylonitrile nanofibers and its application for the photodegradation of methyl blue in aqueous medium. Desalin. Water Treat. **54**(11), 3146–3151 (2015)
63. Cassano, A.E., Alfano, O.M.: Reaction engineering of suspended solid heterogeneous photocatalytic reactors. Catal. Today **58**(2–3), 167–197 (2000)
64. Pareek, V., et al.: Light intensity distribution in heterogenous photocatalytic reactors. Asia-Pac. J. Chem. Eng. **3**(2), 171–201 (2008)
65. Terzian, R., Serpone, N.: Heterogeneous photocatalyzed oxidation of creosote components: mineralization of xylenols by illuminated $TiO_2$ in oxygenated aqueous media. J. Photochem. Photobiol., A **89**(2), 163–175 (1995)
66. Ollis, D.F., Pelizzetti, E., Serpone, N.: Photocatalyzed destruction of water contaminants. Environ. Sci. Technol. **25**(9), 1522–1529 (1991)
67. Herrmann, J.-M.: Heterogeneous photocatalysis: fundamentals and applications to the removal of various types of aqueous pollutants. Catal. Today **53**(1), 115–129 (1999)
68. Kaneco, S., et al.: Optimization of solar photocatalytic degradation conditions of bisphenol A in water using titanium dioxide. J. Photochem. Photobiol., A **163**(3), 419–424 (2004)
69. Pare, B., et al.: ZnO assisted photocatalytic degradation of acridine orange in aqueous solution using visible irradiation. Desalination **232**(1–3), 80–90 (2008)
70. Behnajady, M., et al.: Photocatalytic degradation of an azo dye in a tubular continuous-flow photoreactor with immobilized $TiO_2$ on glass plates. Chem. Eng. J. **127**(1–3), 167–176 (2007)
71. Venkatachalam, N., Palanichamy, M., Murugesan, V.: Sol-gel preparation and characterization of alkaline earth metal doped nano $TiO_2$: efficient photocatalytic degradation of 4-chlorophenol. J. Mol. Catal. A: Chem. **273**(1–2), 177–185 (2007)
72. Chiou, C.-H., Wu, C.-Y., Juang, R.-S.: Influence of operating parameters on photocatalytic degradation of phenol in UV/$TiO_2$ process. Chem. Eng. J. **139**(2), 322–329 (2008)
73. Behnajady, M.A., et al.: Investigation of the effect of heat attachment method parameters at photocatalytic activity of immobilized ZnO nanoparticles on glass plate. Desalination **249**(3), 1371–1376 (2009)
74. Selvam, K., et al.: The influence of inorganic oxidants and metal ions on semiconductor sensitized photodegradation of 4-fluorophenol. Chem. Eng. J. **128**(1), 51–57 (2007)
75. Chiou, C.-H., Wu, C.-Y., Juang, R.-S.: Photocatalytic degradation of phenol and m-nitrophenol using irradiated $TiO_2$ in aqueous solutions. Sep. Purif. Technol. **62**(3), 559–564 (2008)
76. Nikazar, M., Gholivand, K., Mahanpoor, K.: Photocatalytic degradation of azo dye Acid Red 114 in water with $TiO_2$ supported on clinoptilolite as a catalyst. Desalination **219**(1–3), 293–300 (2008)
77. Sahoo, C., Gupta, A., Pal, A.: Photocatalytic degradation of Methyl Red dye in aqueous solutions under UV irradiation using $Ag^+$ doped $TiO_2$. Desalination **181**(1–3), 91–100 (2005)
78. Su, Y., et al.: Photocatalytic degradation of CI Acid Blue 80 in aqueous suspensions of titanium dioxide under sunlight. React. Kinet. Catal. Lett. **98**, 227–240 (2009)
79. Qamar, M., Saquib, M., Muneer, M.: Photocatalytic degradation of two selected dye derivatives, chromotrope 2B and amido black 10B, in aqueous suspensions of titanium dioxide. Dyes Pigm. **65**(1), 1–9 (2005)
80. Daneshvar, N., et al.: Removal of CI Acid Orange 7 from aqueous solution by UV irradiation in the presence of ZnO nanopowder. J. Hazard. Mater. **143**(1–2), 95–101 (2007)
81. Tsai, W.-T., et al.: Photodegradation of bisphenol-A in a batch $TiO_2$ suspension reactor. J. Hazard. Mater. **168**(1), 269–275 (2009)
82. Saquib, M., Muneer, M.: Semiconductor mediated photocatalysed degradation of an anthraquinone dye, Remazol Brilliant Blue R under sunlight and artificial light source. Dyes Pigm. **53**(3), 237–249 (2002)
83. Bahnemann, W., Muneer, M., Haque, M.: Titanium dioxide-mediated photocatalysed degradation of few selected organic pollutants in aqueous suspensions. Catal. Today **124**(3–4), 133–148 (2007)

84. Tariq, M.A., Faisal, M., Muneer, M.: Semiconductor-mediated photocatalysed degradation of two selected azo dye derivatives, amaranth and bismarck brown in aqueous suspension. J. Hazard. Mater. **127**(1–3), 172–179 (2005)
85. Galindo, C., Jacques, P., Kalt, A.: Photodegradation of the aminoazobenzene acid orange 52 by three advanced oxidation processes: UV/$H_2O_2$, UV/$TiO_2$ and VIS/$TiO_2$: comparative mechanistic and kinetic investigations. J. Photochem. Photobiol., A **130**(1), 35–47 (2000)
86. Wenhua, L., et al.: Kinetics of photocatalytic degradation of aniline in water over $TiO_2$ supported on porous nickel. J. Photochem. Photobiol., A **131**(1–3), 125–132 (2000)
87. Alaton, I.A., Balcioglu, I.A.: Photochemical and heterogeneous photocatalytic degradation of waste vinylsulphone dyes: a case study with hydrolyzed Reactive Black 5. J. Photochem. Photobiol., A **141**(2–3), 247–254 (2001)
88. Turchi, C.S., Ollis, D.F.: Photocatalytic degradation of organic water contaminants: mechanisms involving hydroxyl radical attack. J. Catal. **122**(1), 178–192 (1990)
89. Fox, M.A., Dulay, M.T.: Heterogeneous photocatalysis. Chem. Rev. **93**(1), 341–357 (1993)
90. Borker, P., Gaokar, R.D., Fatrekar, A.: Tuning the optical properties of $SnO_2$ by doping $Cu^{2+}$ for enhancing solar irradiated photodegradation of dye effluents and underlying mechanism. Mater. Sci. Eng., B **290**, 116280 (2023)
91. Reddy, M.P., Venugopal, A., Subrahmanyam, M.: Hydroxyapatite-supported Ag–$TiO_2$ as Escherichia coli disinfection photocatalyst. Water Res. **41**(2), 379–386 (2007)
92. Chen, L.-C., Huang, C.-M., Tsai, F.-R.: Characterization and photocatalytic activity of $K^+$-doped $TiO_2$ photocatalysts. J. Mol. Catal. A: Chem. **265**(1–2), 133–140 (2007)
93. Rengaraj, S., et al.: Photocatalytic degradation of methylparathion—an endocrine disruptor by $Bi^{3+}$-doped $TiO_2$. J. Mol. Catal. A: Chem. **247**(1–2), 36–43 (2006)
94. Baran, W., Makowski, A., Wardas, W.: The effect of UV radiation absorption of cationic and anionic dye solutions on their photocatalytic degradation in the presence $TiO_2$. Dyes Pigm. **76**(1), 226–230 (2008)
95. Chakrabarti, S., Dutta, B.K.: Photocatalytic degradation of model textile dyes in wastewater using ZnO as semiconductor catalyst. J. Hazard. Mater. **112**(3), 269–278 (2004)
96. Reza, K.M., Kurny, A., Gulshan, F.: Parameters affecting the photocatalytic degradation of dyes using $TiO_2$: a review. Appl Water Sci **7**, 1569–1578 (2017)
97. Zhu, H., et al.: Effective photocatalytic decolorization of methyl orange utilizing $TiO_2$/ZnO/chitosan nanocomposite films under simulated solar irradiation. Desalination **286**, 41–48 (2012)
98. Alaton, I.A., Balcioglu, I.A., Bahnemann, D.W.: Advanced oxidation of a reactive dyebath effluent: comparison of $O_3$, $H_2O_2$/UV-C and $TiO_2$/UV-A processes. Water Res. **36**(5), 1143–1154 (2002)
99. Borker, P., Salker, A.: Photocatalytic degradation of textile azo dye over $Ce_{1-x}Sn_xO_2$ series. Mater. Sci. Eng., B **133**(1–3), 55–60 (2006)
100. Bahnemann, D., et al.: Aquatic and Surface Photochemistry, p. 261. Lewis, Boca Raton (1994)
101. Tunesi, S., Anderson, M.: Influence of chemisorption on the photodecomposition of salicylic acid and related compounds using suspended titania ceramic membranes. J. Phys. Chem. **95**(8), 3399–3405 (1991)
102. Lachheb, H., et al.: Photocatalytic degradation of various types of dyes (Alizarin S, Crocein Orange G, Methyl Red, Congo Red, Methylene Blue) in water by UV-irradiated titania. Appl. Catal. B **39**(1), 75–90 (2002)
103. Wang, K.-H., et al.: The pH and anion effects on the heterogeneous photocatalytic degradation of o-methylbenzoic acid in $TiO_2$ aqueous suspension. Chemosphere **40**(4), 389–394 (2000)
104. Wei, T.Y., Wan, C.C.: Heterogeneous photocatalytic oxidation of phenol with titanium dioxide powders. Ind. Eng. Chem. Res. **30**(6), 1293–1300 (1991)
105. Hu, C., et al.: Effects of acidity and inorganic ions on the photocatalytic degradation of different azo dyes. Appl. Catal. B **46**(1), 35–47 (2003)
106. Hung, C.-H., Yuan, C.: Reduction of Azo-Dye Via $TiO_2$-photocatalysis. J. Chin. Inst. Environ. Eng. **10**(3), 209–216 (2000)

107. Karimi, L., Zohoori, S., Yazdanshenas, M.E.: Photocatalytic degradation of azo dyes in aqueous solutions under UV irradiation using nano-strontium titanate as the nanophotocatalyst. J. Saudi Chem. Soc. **18**(5), 581–588 (2014)
108. Mozia, S., et al.: Application of anatase-phase $TiO_2$ for decomposition of azo dye in a photocatalytic membrane reactor. Desalination **241**(1–3), 97–105 (2009)
109. Soares, E.T., Lansarin, M.A., Moro, C.C.: A study of process variables for the photocatalytic degradation of rhodamine B. Braz. J. Chem. Eng. **24**, 29–36 (2007)
110. Zhou, S., Ray, A.K.: Kinetic studies for photocatalytic degradation of Eosin B on a thin film of titanium dioxide. Ind. Eng. Chem. Res. **42**(24), 6020–6033 (2003)
111. Mehrotra, K., Yablonsky, G.S., Ray, A.K.: Macro kinetic studies for photocatalytic degradation of benzoic acid in immobilized systems. Chemosphere **60**(10), 1427–1436 (2005)
112. Tian, L., et al.: Visible light enhanced Fe–I–$TiO_2$ photocatalysts for the degradation of gaseous benzene. Atmos. Pollut. Res. **11**(1), 179–185 (2020)
113. Zhu, D., et al.: Highly active and selective catalytic hydrogenation of p-chloronitrobenzene to p-chloroaniline on Pt@ Cu/$TiO_2$. Appl. Surf. Sci. **504**, 144329 (2020)
114. Nyankson, E., et al.: Synthesis and application of Fe-doped $TiO_2$-halloysite nanotubes composite and their potential application in water treatment. Adv. Mater. Sci. Eng. **2019**, 1–15 (2019)
115. Bhardwaj, S., et al.: Photodeposition time dependant growth, size and photoactivity of Ag and Cu deposited $TiO_2$ nanocatalyst under solar irradiation. Sol. Energy **194**, 618–627 (2019)
116. Ghanbari, S., Givianrad, M.H., Aberoomand Azar, P.: Synthesis and characterization of visible light driven N—Fe-codoped $TiO_2$/$SiO_2$ for simultaneous photoremoval of Cr (VI) and azo dyes in a novel fixed bed continuous flow photoreactor. Can. J. Chem. Eng. **98**(3), 705–716 (2020)
117. Ahadi, S., Moalej, N.S., Sheibani, S.: Characteristics and photocatalytic behavior of Fe and Cu doped $TiO_2$ prepared by combined sol-gel and mechanical alloying. Solid State Sci. **96**, 105975 (2019)
118. Zhang, J., et al.: Photocatalytic removal of chromium (VI) and sulfite using transition metal (Cu, Fe, Zn) doped $TiO_2$ driven by visible light: feasibility, mechanism and kinetics. J. Ind. Eng. Chem. **80**, 23–32 (2019)
119. Khan, M.M., et al.: Role of Fe doping in tuning photocatalytic and photoelectrochemical properties of $TiO_2$ for photodegradation of methylene blue. Opt. Laser Technol. **118**, 170–178 (2019)
120. Stucchi, M., et al.: Nano-$MnO_2$ decoration of $TiO_2$ microparticles to promote gaseous ethanol visible photoremoval. Nanomaterials **8**(9), 686 (2018)
121. Choi, W., Termin, A., Hoffmann, M.R.: The role of metal ion dopants in quantum-sized $TiO_2$: correlation between photoreactivity and charge carrier recombination dynamics. J. Phys. Chem. **98**(51), 13669–13679 (2002)
122. Du, X., et al.: Optimization of the Ge/$TiO_2$ catalyst for the degradation of ciprofloxacin by the response surface methodology. Zhongguo Kangshengsu Zazhi **44**, 750–757 (2019)
123. Crişan, M., et al.: Iron doped $TiO_2$ films and their photoactivity in nitrobenzene removal from water. Appl. Surf. Sci. **455**, 201–215 (2018)
124. Gnanasekaran, L., et al.: Intermediate state created by dopant ions (Mn, Co and Zr) into $TiO_2$ nanoparticles for degradation of dyes under visible light. J. Mol. Liq. **223**, 652–659 (2016)
125. Huang, J.-G., et al.: Synthesis and photocatalytic activity of Mo-doped $TiO_2$ nanoparticles. J. Spectro. **2015** (2015)
126. Li, R., Li, T., Zhou, Q.: Impact of titanium dioxide ($TiO_2$) modification on its application to pollution treatment—a review. Catalysts **10**(7), 804 (2020)
127. Qin, Y., et al.: Nitrogen-doped hydrogenated $TiO_2$ modified with CdS nanorods with enhanced optical absorption, charge separation and photocatalytic hydrogen evolution. Chem. Eng. J. **384**, 123275 (2020)
128. Ratova, M., et al.: Synergistic effect of doping with nitrogen and molybdenum on the photocatalytic properties of thin titania films. Vacuum **114**, 205–212 (2015)

129. Kovalevskiy, N., et al.: Synergistic effect of polychromatic radiation on visible light activity of N-doped $TiO_2$ photocatalyst. Catal. Commun. **134**, 105841 (2020)
130. Fang, W., Xing, M., Zhang, J.: Modifications on reduced titanium dioxide photocatalysts: a review. J. Photochem. Photobiol., C **32**, 21–39 (2017)
131. Reda, S., Khairy, M., Mousa, M.: Photocatalytic activity of nitrogen and copper doped $TiO_2$ nanoparticles prepared by microwave-assisted sol-gel process. Arab. J. Chem. **13**(1), 86–95 (2020)
132. Jin, X., et al.: Photocatalytic degradation of norfloxacin using N-doped $TiO_2$: optimization, mechanism, identification of intermediates and toxicity evaluation. Chemosphere **237**, 124433 (2019)
133. Ji, L., et al.: Synthesis of carbon-and nitrogen-doped $TiO_2$/carbon composite fibers by a surface-hydrolyzed PAN fiber and their photocatalytic property. J. Mater. Sci. **55**(6), 2471–2481 (2020)
134. Asahi, R., et al.: Visible-light photocatalysis in nitrogen-doped titanium oxides. Science **293**(5528), 269–271 (2001)
135. Li, H., et al.: A systematic study on visible-light N-doped $TiO_2$ photocatalyst obtained from ethylenediamine by sol–gel method. Appl. Surf. Sci. **344**, 112–118 (2015)
136. Jyothi, M., et al.: Novel hydrothermal method for effective doping of N and F into nano Titania for both, energy and environmental applications. Mater. Res. Bull. **74**, 478–484 (2016)
137. Rahbar, M., et al.: S, N co-doped carbon quantum dots/$TiO_2$ nanocomposite as highly efficient visible light photocatalyst. Nanotechnology **30**(50), 505702 (2019)
138. Liu, D., et al.: Fabricating I doped $TiO_2$ photoelectrode for the degradation of diclofenac: Performance and mechanism study. Chem. Eng. J. **369**, 968–978 (2019)
139. John, B., et al.: A review on carbon quantum dot based semiconductor photocatalysts for the abatement of refractory pollutants. Chem. Phys. Chem: Eur. J. Chem. Phys. Phys. Chem. (2022)
140. Torres-Pinto, A., et al.: Photo-Fenton degradation assisted by in situ generation of hydrogen peroxide using a carbon nitride photocatalyst. J. Water Process Eng. **37**, 101467 (2020)
141. Qu, D., et al.: Highly luminescent S, N co-doped graphene quantum dots with broad visible absorption bands for visible light photocatalysts. Nanoscale **5**(24), 12272–12277 (2013)
142. Cai, A., et al.: Hierarchical ZnO/S, N: GQD composites: biotemplated synthesis and enhanced visible-light-driven photocatalytic activity. Appl. Surf. Sci. **391**, 484–490 (2017)
143. Serp, P., Corrias, M., Kalck, P.: Carbon nanotubes and nanofibers in catalysis. Appl. Catal. A **253**(2), 337–358 (2003)
144. Kokubo, K., et al.: Facile synthesis of highly water-soluble fullerenes more than half-covered by hydroxyl groups. ACS Nano **2**(2), 327–333 (2008)
145. Yan, Y., et al.: Carbon nanotube catalysts: recent advances in synthesis, characterization and applications. Chem. Soc. Rev. **44**(10), 3295–3346 (2015)
146. Zhang, J., et al.: Recent progress on carbon-nanotube-based materials for photocatalytic applications: a review. Solar RRL **6**(9), 2200243 (2022)
147. Sarkodie, B., et al.: Photocatalytic degradation of dyes by novel electrospun nanofibers: a review. Chemosphere **313**, 137654 (2023)
148. Mohapatra, L., Cheon, D., Yoo, S.H.: Carbon-based nanomaterials for catalytic wastewater treatment: a review. Molecules **28**(4), 1805 (2023)
149. Liang, S., et al.: Fabrication of graphdiyne and its analogues for photocatalytic application. EcoMat **5**(3), e12297 (2023)
150. Wang, S., et al.: A novel and highly efficient photocatalyst based on P25–graphdiyne nanocomposite. Small **8**(2), 265–271 (2012)
151. Hou, H., Shao, G., Yang, W.: Recent advances in $gC_3N_4$-based photocatalysts incorporated by MXenes and their derivatives. J. Mater. Chem. A **9**(24), 13722–13745 (2021)
152. Patnaik, S., Martha, S., Parida, K.: An overview of the structural, textural and morphological modulations of $gC_3N_4$ towards photocatalytic hydrogen production. RSC Adv. **6**(52), 46929–46951 (2016)

153. Liu, W., Wang, B., Zhang, M.: Effect of process parameters on the microstructure and performance of $TiO_2$-loaded activated carbon. ACS Omega **6**(50), 35076–35092 (2021)
154. Yang, Q., et al.: Nanostructured heterogeneous photo-catalysts for hydrogen production and water splitting: a comprehensive insight. Appl. Mater. Today **17**, 159–182 (2019)
155. Patial, S., et al.: A review on carbon quantum dots modified g-$C_3N_4$-based photocatalysts and potential application in wastewater treatment. Appl. Sci. **12**(21), 11286 (2022)
156. Fang, S., et al.: Effect of carbon-dots modification on the structure and photocatalytic activity of g-$C_3N_4$. Appl. Catal. B **185**, 225–232 (2016)

# Chapter 5
# Antibacterial Application

## 5.1 Physical/Mechanical Destruction

All bacteria need a cell wall to retain shape, control osmotic pressure, resist mechanical stress, and combat infection. Disruption of the bacterial outer membrane or cell wall can lead to dysfunction and leaking of cytoplasm, resulting in bacteriostatic and bactericidal effects (Fig. 5.1a). This route is one of the most prevalent antibacterial pathways in CBNMs of various dimensions.

Almost all forms of 0D CBNMs have the ability to cause mechanical harm to the external membrane or cell wall of bacteria. Carboxylated nanodiamonds exhibit an antibacterial effect by causing damage to the cell wall of *E. coli.* The impact is caused by the binding of the CBNM to the cell wall surface, resulting in changes to the protein structure of the bacterial cell wall [1]. Furthermore, positively charged CDs like spermidine CDs can interact with porins, peptidoglycans, and proteins present in the cell membrane. This bonding is formed by electrostatic contact and hydrogen bonding.

As a result, the cell membrane becomes destabilized in a synergistic manner, and the synthesis of the membrane is inhibited [2]. Similarly, an extraordinary breakdown of the membrane was observed in model systems of enormous unilamellar vesicles after being exposed to $C_{60}$. This disruption was identified by the use of fluorescence imaging (Fig. 5.1b) [3]. By identifying the leaking of intracellular components such as DNA and RNA, as well as by assessing the integrity of the cell membrane, it is possible to determine the amount of the damage that has been caused to the cell wall or cell membrane as a result of being exposed to 1D CBNMs such as CNTs.

In order to eliminate the influence of oxidative stress and metal toxicity caused by CNTs, we utilize pure SWCNTs to examine their antibacterial properties and understand the underlying process. The data obtained indicate that the antibacterial effect of SWCNTs may be attributed to their ability to penetrate the bacterial cell membrane. Later on, it was discovered, individually distributed SWCNTs functioned as "nanodarts" to physically breach a bacterial cell's outer membrane, as seen by



M. Ikram et al., *Carbon-Based Nanomaterials for Environmental Applications*, Engineering Materials, https://doi.org/10.1007/978-3-031-59390-1_5

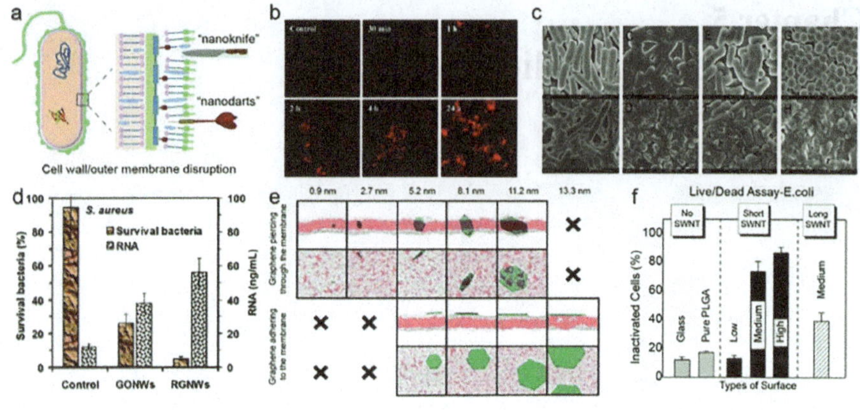


**Fig. 5.1** **a** Physical/mechanical destruction triggered by CBNMs on bacteria. Reproduced with permission from [11], Copyright 2018, Advanced materials. **b** Morphological changes of RhB-labeled $GUVs^+$ after exposure to $C_{60}$ NPs in 0.1 m glucose. Reproduced with permission from [3] Copyright 2018, The Royal Society of Chemistry. **c** SEM images displaying loss of membrane integrity after SWCNTs' treatment. Reproduced with permission from [12], Copyright 2009, ACS. **d** Cytotoxicity of GONWs and RGNWs to *S. aureus*, and concentrations of RNA in PBS of *S. aureus* bacteria exposed to nanowalls. Reproduced with permission from [6] Copyright 2010, ACS. **e** Theoretical simulation to reveal the relationship between the mechanical destruction action and the size of a Gr nanosheet Reproduced with permission [13] Copyright 2015, ACS. **f** Fraction of inactivated *E. coli* on SWNT-PLGA films (%) for various SWNT lengths and concentrations. Reproduced with permission [10], Copyright 2010, The Royal Society of Chemistry

SEM pictures (Fig. 5.1c). [4]. In addition, both SWCNTs and MWCNTs exhibit a "needle-like" mechanism of action on the bacterial cell wall, leading to the inhibition of bacterial growth by membrane depolarization [5].

GO nanowalls (GONWs) and rGO nanowalls (RGNWs), which are 2D CBNMs, have been found to exhibit a blade/knife-like mechanism that leads to the destruction of bacterial membranes. Verification was conducted by quantifying the release of bacterial internal components, such as RNA, into phosphate buffer saline following interaction with GONWs/RGNWs (Fig. 5.1d) [6]. A new investigation confirmed that the membrane rupture induced by GO is increased when the orientation of GO nanosheets changes from horizontally oriented to perpendicular. This is due to the increased density of edges in the vertical orientation [7]. Gram-positive (G +ve) microbes like Staphylococcus aureus, without an outer membrane, are more vulnerable to direct contact with the sharp edges of RGNWs than Gram-negative (G –ve) microbes. Theoretical simulations further support the mechanical damaging effect and demonstrate its association with the size of the Gr nanosheet [8]. The results suggest that tiny hydrophobic Gr sheets can readily enter the phospholipid membrane, leading to mechanical harm. Intermediate-sized sheets can only enter the membrane in precise geometric layouts, while bigger sheets are unlikely to trigger mechanical harm as they stick to the lipid bilayer instead of entering it (Fig. 5.1e).

Certain CBNMs of varying dimensionalities have the ability to cause mechanical harm to the external membrane or cell wall of bacteria. Nevertheless, the degree of mechanical harm differs among CBNMs with distinct dimensionalities, likely because of the variation in the nature of contact interaction amid the CBNMs and the cell wall of microbes. These interactions may be classified as exchanges occurring at the 0D, 1D, and 2D interfaces between CBNMs and bacteria, corresponding to point, line, and plane connections, respectively. Mechanical damage impact may be managed by several factors, such as the size, length, and distribution of the CBNMs, along with the bacterial culture parameters like the agitation speed. Using CNTs as a specific example, the findings indicate that CNTs with smaller diameters tend to come into touch with bacterial cells near the end of the tube, whereas CNTs with larger diameters mostly interact with cells through their side walls. Therefore, CNTs with smaller dimensions are more likely to have a significant needle-like impact on the bacterial cell wall compared to those with a greater diameter [9]. When the weight proportion of SWCNTs is supplied, shorter CNTs have a larger density of tube ends compared to longer ones. This increases the likelihood of mechanical distraction caused by a tube end, leading to antimicrobial efficacy (Fig. 5.1f) [10]. In contrast to aggregated CNTs, well-dispersed CNTs function as "nanodarts" and have the ability to significantly increase the piercing impact on the bacterial cell membrane. At increased shaking speeds, the movement of both CNTs and bacterial cells in the fluid increases. This leads to more frequent and intense collisions between the CNTs and bacteria, causing more severe damage to the cell membranes. Hence, CNTs possessing a smaller diameter, shorter length, improved dispersion, and increased shaking speed would result in a more efficient antibacterial impact by causing mechanical harm to the cell wall or cell membrane [4].

## 5.2 Oxidative Stress

The mechanism of CBNMs induced oxidative stress, which has similar physicochemical features with CNTs, is widely accepted [14–16]. Oxidative stress hampers bacterial metabolism and disrupts vital cellular processes, ultimately resulting in cellular inactivation or even cell demise. Typically, oxidative stress primarily arises through either a route that relies on ROS or a pathway that does not depend on ROS. The former is caused by the excessive buildup of intracellular ROS, including $H_2O_2$, $O_2^{\cdot-}$, $OH^{\cdot}$, or singlet molecular oxygen ($^1O_2$). Extensive evidence shows that ROS triggers intracellular protein inactivation, lipid peroxidation, mitochondrial malfunction, and progressive degradation of the cell membrane, leading to necrosis/apoptosis and ultimately cell demise. The latter refers to the process of causing disturbance or oxidation in a crucial cellular structure or component without the generation of ROS. This can occur when there is a transfer of charge from the cellular membrane to graphene, with graphene acting as an electron pump [17].

### 5.2.1 ROS-Dependent Oxidative Stress

Graphene materials (GMs) can facilitate the production of ROS by the absorption of $O_2$ on the defective areas and edges of the GMs. This is then followed by the reduction of $O_2$ by different biological enzymes that have a reducing effect, such as glutathione (GSH) [18–20]. Glutathione (GSH) is a significant molecule with antioxidant properties that can undergo oxidation to become glutathione disulfide (GSSG) when ROS are present. Therefore, GSH has the potential to function as an indication of the redox status inside cells, and its reduction suggests the harmful impact of oxidative stress on bacteria [21–24]. Besides GSH, antioxidants such as *N*-acetylcysteine and α-tocopherol as well as oxidant-sensitive dyes like dichlorodihydrofluorescein diacetate (DCFH-DA) might indicate ROS production [22, 24–28]. Chen et al. observed that the levels of ROS increased in a way that depended on the concentration of graphene when *Xanthomonas oryzae pv Oryzae* was exposed to graphene. This was determined by monitoring the ROS levels using DCFH-DA, as shown in Fig. 5.2a-A. Additionally, they noted that the oxidation of DCFH-DA cannot occur only through the presence of GO/rGO, suggesting that the generation of ROS is responsible for inducing oxidative stress. Furthermore, a study has demonstrated that the antibacterial properties of graphene are strongly linked to the disruption of the mitochondrial membrane potential, which is a crucial step in the buildup of ROS within cells. Specifically, it was found that pristine graphene can cause a gradual decrease in the mitochondrial membrane potential of murine RAW 264.7 macrophages, and this effect is dependent on the dosage of graphene administered [27].

In addition, Kim's research team discovered that the levels of ROS in P. aeruginosa cells treated with GO and rGO were 3.8 and 2.7 times higher, respectively, compared to the levels in control cells [22]. Consistently, prior exposure of the cells to GSH or NAC, which are antioxidants, markedly decreased the amounts of ROS produced by GO/rGO. This confirms that cell death is caused by the generation of ROS. They demonstrated that the generation of $O_2^{\cdot-}$ by GO and rGO dispersions was, respectively, two and one times more than in control cells, providing direct evidence of the harmful effects of ROS on *E. coli* (Fig. 5.2a-B) [22]. These findings indicate that oxidative stress plays a significant role in the antibacterial effects of GMs. The presence of oxygen-containing functional groups and the density of defects on the surface of GMs significantly contribute to the production of ROS. In their study, Musico et al. discovered that a membrane filter changed with PVK-GO resulted in a higher depletion of GSH compared to a filter modified with PVK-G [23]. This suggests that the increased formation of RO would cause more oxidative stress to the cells of *E. coli* and *B. subtilis*. It was shown that a higher quantity of oxygen-containing functional groups (such as –COOH and –OH) on the surface of GO can enhance the generation of ROS and augment its antibacterial characteristics. Concurrently, the ability of GO sheets to oxidize GSH rose from 49 to 71% as the sheet area reduced from 0.65 to 0.01 $\mu m^2$. This can be attributed to the higher occurrence of defects in smaller GO sheets, as seen in Fig. 5.2a-C [18]. Furthermore, cells that are exposed to GO often exhibit a greater level of oxidative stress compared

(a)

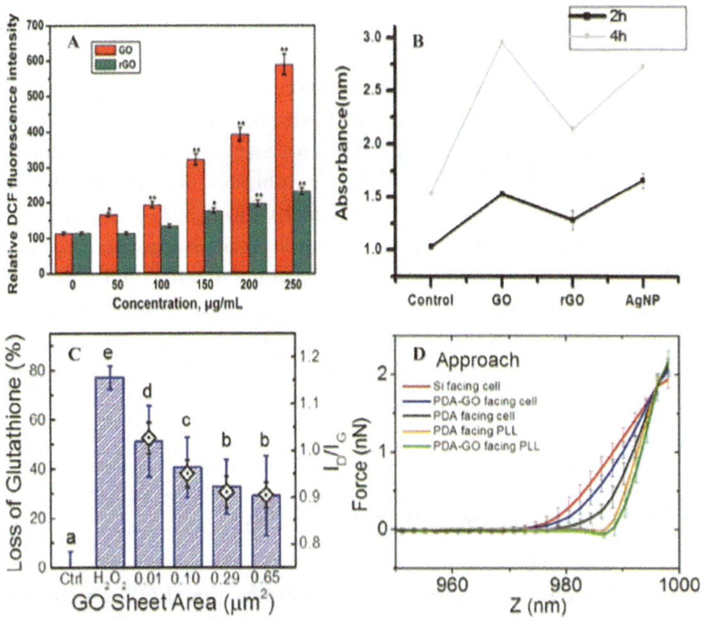


**Fig. 5.2** **a** ROS generation after incubation of bacteria suspensions in DI water (A), production of $O_2^{\cdot -}$ by GO and rGO dispersions (B), in vitro GSH oxidation by GO sheets of different areas, (C) plots of the force as a function of piezo-position (Z) during cantilever approach to *E. coli* cells, (D), Reproduced with permission from [29], Copyright 2016, ACS

to cells exposed to rGO. This is likely due to the high defect density of GO and its exceptional ability to disperse [29].

Lipid peroxidation is a process in which lipid molecules undergo an array of radical chain events due to the oxidation caused by ROS. This pathway is significant because of the strong connection between lipid peroxidation and the cell membranes of GMs [30, 31]. Lipid peroxidation can generate lipid peroxide radicals, which then spread oxidative damage throughout cell membranes. Krishnamoorthy et al. utilized an ultrasonic irradiation-induced lipid peroxidation experiment to assess the activities of graphene nanosheets that are influenced by free radicals [31]. Through this approach, they identified three distinct reaction products (conjugated dienes, lipid hydroperoxides, and malondialdehydes) that were generated during various phases of lipid peroxidation facilitated by free radicals (such as $OH^{\cdot}$). In addition, Perreault et al. employed the lipid-soluble antioxidant α-tocopherol, which is known for its significant antioxidant properties in preventing lipid peroxidation by neutralizing free radicals [32]. This was done to validate the involvement of oxidative stress in the inactivation of bacteria generated by GO. The researchers noticed that when *E. coli* cells were exposed to α-tocopherol before incubation, the antibacterial effects of all sizes of GO sheets were diminished, with bigger sheets being affected more significantly. These findings provide partial evidence for the role of free radicals in the reaction pathway, thereby confirming the participation of ROS in the antibacterial activities.

An essential aspect in assessing the effectiveness of antibacterial agents is comprehending the interplay between genetically GMs and the membranes of bacterial cells. Castrillon et al. employed atomic force microscopy (AFM) to investigate the interaction between GO and *E. coli* cellular membranes, aiming to gain insight into the mechanism of GO-induced bacterial inactivation [33]. Using a polydopamine-assisted functionalized AFM probe, the researchers measured the force of interaction between the cellular membrane and GO. They found that the interactions between the cell and GO are mostly repulsive, indicating that physical interactions like membrane penetration are not necessary (Fig. 5.2a-D). In addition, they showcased the involvement of oxidative pathways in the antibacterial action of GO by presenting evidence of the oxidation of GSH by GO [29].

### *5.2.2 ROS-Independent Oxidative Stress*

Although experimental data clearly demonstrates that ROS-dependent oxidative stress plays a major role in the antibacterial activity of CNMs, several research groups have shown that CNMs may also elicit antibacterial effects through ROS-independent oxidative stress. This result is derived from the finding that there is an absence of ROS generation or ROS-induced harm after administering $^{n}C_{60}$. Furthermore, the bactericidal efficacy of $^{n}C_{60}$ is unaffected by both light and oxygen, which are crucial factors for the generation of ROS [34, 35]. A recent investigation has validated this idea, indicating that $^{n}C_{60}$, rather than ROS, directly interacts with the cell membrane and disrupts the electron transport process involved in energy transduction during respiration [36].

Furthermore, ample data has been shown to substantiate the ROS-independent antibacterial mechanism in materials made from graphene. When employing the 2,3-bis (2-methoxy-4-nitro-5-sulfophenyl)-2*H*-tetrazolium-5-carboxanilide technique to quantify the existence of superoxide, minimal amounts of $O_2^{\cdot -}$ (other reactive oxygen species such as singlet oxygen and hydroxyl, which can be produced from $O_2^{\cdot -}$) were seen in the GO and RGO suspensions. Results from Ellman's experiment demonstrate that different graphene-based materials have varying oxidation capabilities toward GSH. This indirectly shows that graphene-based materials are capable of mediating oxidative stress that is independent of $O_2^{\cdot -}$ [37]. Metallic single-walled SWNTs exhibit comparable cytotoxicity to ROS-independent mechanisms, such as GSH oxidation [38]. Moreover, it has been suggested that the antibacterial properties of graphene might be ascribed to the disruption of electron transport in the membrane respiratory chain, which leads to oxidative stress independent of reactive oxygen species (ROS), rather than damage caused by ROS [39]. Additional findings indicate that both the graphene@Cu and graphene@Ge surfaces exhibit potent antibacterial properties due to the electron transfer from the microbial membrane to the graphene film, which is then transmitted to the underlying conducting Cu or semiconducting Ge substrate. On the other hand, the antibacterial effectiveness of the graphene@$SiO_2$ surface is minimal due to the inability of electrons to move from the microbial membrane to the underlying insulator $SiO_2$ substrate via graphene. This process enhances the transfer of electrons from bacterial antioxidant biomolecules to GO, leading to the reduction of GO [40]. Collectively, our findings validate that materials based on graphene can also trigger a bactericidal impact by means of oxidative stress that is not dependent on ROS. GO has a higher abundance of surface oxygen carriers compared to RGO, which enhances its ability to assist oxidation processes. As a result, GO demonstrates much superior antibacterial efficacy [41–43]. Despite being combined with poly(*N*-vinylcarbazole) (PVK) to create nanocomposite filters, the PVK-GO composite exhibits superior antibacterial activity compared to PVK-RGO/graphene [44]. Furthermore, when the surface area of the GO sheet is reduced from 0.65 to 0.01 $\mu m^2$, there is a fourfold enhancement in the antimicrobial activity. This improvement can be attributed to the greater concentration of edge-type defects in the smaller sheets, as evidenced by the higher $I_D/I_G$ ratio in the Raman spectra. These defects act as sites for oxygen adsorption. The results suggest that increasing the concentration of oxygen-containing active sites in graphene-based nanomaterials enhances their ability to combat bacteria through oxidative mechanisms that are not ROS [45].

## 5.3 Photothermal Effect

During antibacterial photothermal treatment (PTT), the bacterial membrane is initially damaged by the CDs by the absorption of photons and subsequent internal conversion, resulting in an increase in temperature. CDs adhere to the bacterial membrane primarily through electrostatic interactions and transmit heat to bacteria [46]. Consequently, the bacteria becomes susceptible to high temperatures, facilitating the entry of CDs that intensify cell harm by triggering the production of ROS. Thus, it is customary for PTT to collaborate with Antibacterial Photodynamic Therapy (APDT) in order to enhance the bactericidal action. The addition of an antibiotic or antibacterial agent, such as quaternary ammonium, enhances the initial damage induced by the CDs. This agent amplifies the damage to the bacterial membrane, resulting in a more pronounced effect compared to the CDs alone [47]. The photothermal effect is efficacious in both G +ve and G −ve bacteria as a result of the thermal energy generated by the CDs, which impacts their distinct structures of peptidoglycan, phospholipids, and lipopolysaccharides (LPS). Consequently, CDs in antibacterial PTT have characteristics similar to those of wide-ranging antibiotics. Nevertheless, the absence of irradiation may diminish the efficacy of the antibacterial impact on G +ve bacteria, mostly because of the presence of numerous layers of peptidoglycan. The efficacy of GQDs-AgNPs nanohybrids at a concentration of 2 $\mu g{\cdot}mL^{-1}$ has been demonstrated against G +ve *S. aureus* and G −ve *E. coli* bacterial strains. However, complete suppression of bacterial strains may be accomplished by exposing them to red light (808 nm) from a 2-W·$cm^{-2}$ source for 10 min, along with the same concentration of GQDs-AgNPs [48].

An important factor in antibacterial PTT is the photothermal efficiency of carbon dots (conversion of light energy into heat). During dispersion in a liquid medium, these entities function as sources of heat emission. Depending on the efficiency of the medium in distributing this thermal energy, it is possible to achieve elevated temperatures that might result in harm to healthy tissue. The temperature achieved in in vivo and in vitro applications depends on the photothermal efficiency, the concentration of CDs, the duration of irradiation, the dispersal medium, and the light dosage administered. Figure 5.3a illustrates the relationship between temperature and the scattering of CDs (containing Fe), concentration, and duration of irradiation. The primary determinant for regulating temperature is the concentration of CDs. However, in in vivo applications, temperature measurements are typically performed in real time to prevent undesired harm to tissues. The efficacy of this method is also contingent upon the severity of the illness being treated. CDs containing Fe or Ag nanoparticles (AgNPs) can exhibit peroxidase-like behavior, similar to an enzyme called peroxidase (POD) [48]. These CDs interact with H2O2, leading to an enhancement in their antibacterial activity (99.85% inhibition of *E. coli*) and facilitating the healing process (refer to Fig. 5.3b and c).

Hybrid nanomaterials, namely, chitosan (CS)-nanohydroxyapatite (nHA) scaffolds doped with CDs (CS-nHA-CDs), enable the treatment of bone infections via antibacterial PTT. CS-nHA-CD scaffolds serve as a foundational material for the development of new bone tissue, while also possessing antibacterial properties. The antibacterial scaffolds effectively suppress *E. coli* and *S. aureus* bacteria by 97% and 99%, respectively, when exposed to red light irradiation (808 nm, 1 W·$cm^{-2}$) for 10 min. Healthy tissues attenuate the amount of light that reaches antibacterial scaffolds. Hence, the regulation of temperature and the thermal efficiency of these materials is crucial [48].

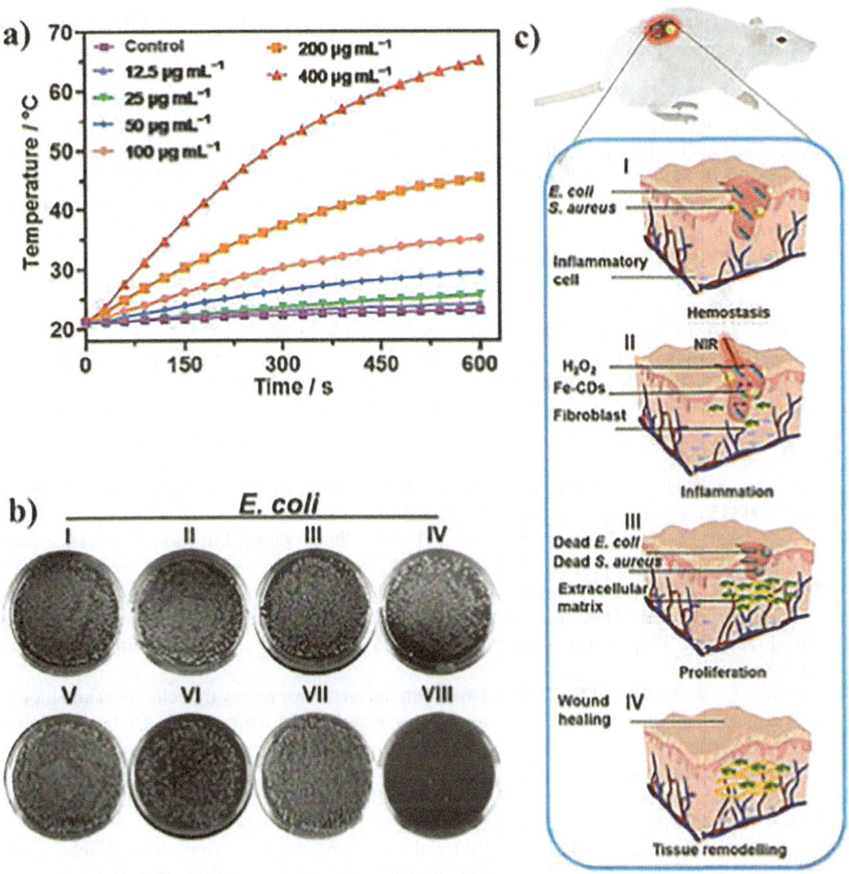


**Fig. 5.3** **a** Thermal study of CDs doped with Fe for different concentrations as a function of irradiation time. **b** The antibacterial activity of Fe-CDs for a strain of *E. coli*, with an irradiation of 2 W·cm$^{-2}$ at 808 nm, (I) control, (II) Fe-CDs, (III) $H_2O_2$, (IV) Fe-CDs + $H_2O_2$, (V) NIR, (VI) Fe-CDs + NIR, (VII) $H_2O_2$ + NIR, (VIII) Fe-CDs + $H_2O_2$ + NIR. **c** Antibacterial PTT in vivo, with wound-healing effect. Reproduced with permission from [49]. Copyright 2023, Biomaterial and tissue engineering

## References


1. Chatterjee, A., Perevedentseva, E., Jani, M., et al.: Antibacterial effect of ultrafine nanodiamond against gram-negative bacteria *Escherichia coli*. J. Biomed. Opt. **20**, 51014 (2014). https://doi.org/10.1117/1.jbo.20.5.051014
2. Jian, H.J., Wu, R.S., Lin, T.Y., et al.: Super-cationic carbon quantum dots synthesized from spermidine as an eye drop formulation for topical treatment of bacterial keratitis. ACS Nano **11**, 6703–6716 (2017). https://doi.org/10.1021/acsnano.7b01023
3. Zeng, Y., Wang, Q., Zhang, Q., Jiang, W.: Quantification of C60-induced membrane disruption using a quartz crystal microbalance. RSC Adv. **8**, 9841–9849 (2018). https://doi.org/10.1039/c7ra13690k

4. Kang, S., Herzberg, M., Rodrigues, D.F., Elimelech, M.: Antibacterial effects of carbon nanotubes: size does matter! Langmuir **24**, 6409–6413 (2008). https://doi.org/10.1021/la800951v
5. Chen, H., Wang, B., Gao, D., et al.: Broad-spectrum antibacterial activity of carbon nanotubes to human gut bacteria. Small **9**, 2735–2746 (2013). https://doi.org/10.1002/smll.201202792
6. Akhavan, O., Ghaderi, E.: Toxicity of graphene and graphene oxide nanowalls against bacteria. ACS Nano **4**, 5731–5736 (2010). https://doi.org/10.1021/nn101390x
7. Lu, X., Feng, X., Werber, J.R., et al.: Enhanced antibacterial activity through the controlled alignment of graphene oxide nanosheets. Proc. Natl. Acad. Sci. U. S. A. **114**, E9793–E9801 (2017). https://doi.org/10.1073/pnas.1710996114
8. Li, Y., Yuan, H., Von Dem Bussche, A., et al.: Graphene microsheets enter cells through spontaneous membrane penetration at edge asperities and corner sites. Proc. Natl. Acad. Sci. U. S. A. **110**, 12295–12300 (2013). https://doi.org/10.1073/pnas.1222276110
9. Arias, L.R., Yang, L.: Inactivation of bacterial pathogens by carbon nanotubes in suspensions. Langmuir **25**, 3003–3012 (2009). https://doi.org/10.1021/la802769m
10. Aslan, S., Loebick, C.Z., Kang, S., et al.: Antimicrobial biomaterials based on carbon nanotubes dispersed in poly(lactic-co-glycolic acid). Nanoscale **2**, 1789–1794 (2010). https://doi.org/10.1039/c0nr00329h
11. Xin, Q., Shah, H., Nawaz, A., et al.: Antibacterial carbon-based nanomaterials. Adv. Mater. **31**, 1804838 (2019). https://doi.org/10.1002/adma.201804838
12. Liu, S., Wei, L., Hao, L., et al.: Sharper and faster "Nano darts" kill more bacteria: a study of antibacterial activity of individually dispersed pristine single-walled carbon nanotube. ACS Nano **3**, 3891–3902 (2009). https://doi.org/10.1021/nn901252r
13. Dallavalle, M., Calvaresi, M., Bottoni, A., et al.: Graphene can wreak havoc with cell membranes. ACS Appl. Mater. Interfaces **7**, 4406–4414 (2015). https://doi.org/10.1021/am508938u
14. Smith, S.C., Rodrigues, D.F.: Carbon-based nanomaterials for removal of chemical and biological contaminants from water: a review of mechanisms and applications. Carbon N. Y. **91**, 122–143 (2015)
15. Lewinski, N., Colvin, V., Drezek, R.: Cytotoxicity of nanopartides. Small **4**, 26–49 (2008)
16. Liu, Y., Zhao, Y., Sun, B., Chen, C.: Understanding the toxicity of carbon nanotubes. Acc. Chem. Res. **46**, 702–713 (2013). https://doi.org/10.1021/ar300028m
17. Li, J., Wang, G., Zhu, H., et al.: Antibacterial activity of large-area monolayer graphene film manipulated by charge transfer. Sci. Rep. **4** (2014). https://doi.org/10.1038/srep04359
18. Perreault, F., De Faria, A.F., Nejati, S., Elimelech, M.: Antimicrobial properties of graphene oxide nanosheets: why size matters. ACS Nano **9**, 7226–7236 (2015). https://doi.org/10.1021/acsnano.5b02067
19. Palmieri, V., Bugli, F., Lauriola, M.C., et al.: Bacteria meet graphene: modulation of graphene oxide nanosheet interaction with human pathogens for effective antimicrobial therapy. ACS Biomater. Sci. Eng. **3**, 619–627 (2017). https://doi.org/10.1021/acsbiomaterials.6b00812
20. Qiu, J., Geng, H., Wang, D., et al.: Layer-number dependent antibacterial and osteogenic behaviors of graphene oxide electrophoretic deposited on titanium. ACS Appl. Mater. Interfaces **9**, 12253–12263 (2017). https://doi.org/10.1021/acsami.7b00314
21. Zhu, J., Xu, M., Gao, M., et al.: Graphene oxide induced perturbation to plasma membrane and cytoskeletal meshwork sensitize cancer cells to chemotherapeutic agents. ACS Nano **11**, 2637–2651 (2017). https://doi.org/10.1021/acsnano.6b07311
22. Lahiani, M.H., Gokulan, K., Williams, K., Khare, S.: Impact of pristine graphene on intestinal microbiota assessed using a bioreactor-rotary cell culture system. ACS Appl. Mater. Interfaces **11**, 25708–25719 (2019). https://doi.org/10.1021/acsami.9b07635
23. Yang, Y., Wu, X., He, C., et al.: Metal-organic framework/ag-based hybrid nanoagents for rapid and synergistic bacterial eradication. ACS Appl. Mater. Interfaces **12**, 13698–13708 (2020). https://doi.org/10.1021/acsami.0c01666
24. Guo, Z., Zhang, P., Xie, C., et al.: Defining the surface oxygen threshold that switches the interaction mode of graphene oxide with bacteria. ACS Nano **17**, 6350–6361 (2023). https://doi.org/10.1021/acsnano.2c10961

25. Cheng, C., Li, S., Thomas, A., et al.: Functional graphene nanomaterials based architectures: biointeractions, fabrications, and emerging biological applications. Chem. Rev. **117**, 1826–1914 (2017)
26. Dallavalle, M., Bottoni, A., Calvaresi, M., Zerbetto, F.: Functionalization pattern of graphene oxide sheets controls entry or produces lipid turmoil in phospholipid membranes. ACS Appl. Mater. Interfaces **10**, 15487–15493 (2018). https://doi.org/10.1021/acsami.8b03224
27. Guo, H., Li, T., Cao, X., et al.: Self-sterilized flexible single-electrode triboelectric nanogenerator for energy harvesting and dynamic force sensing. ACS Nano **11**, 856–864 (2017). https://doi.org/10.1021/acsnano.6b07389
28. Qin, Y., Hu, Y., Koehler, S., et al.: Ultrafast nanofiltration through large-area single-layered graphene membranes. ACS Appl. Mater. Interfaces **9**, 9239–9244 (2017). https://doi.org/10.1021/acsami.7b00504
29. Zou, X., Zhang, L., Wang, Z., Luo, Y.: Mechanisms of the antimicrobial activities of graphene materials. J. Am. Chem. Soc. **138**, 2064–2077 (2016). https://doi.org/10.1021/jacs.5b11411
30. Zhao, R., Lv, M., Li, Y., et al.: Stable nanocomposite based on PEGylated and silver nanoparticles loaded graphene oxide for long-term antibacterial activity. ACS Appl. Mater. Interfaces **9**, 15328–15341 (2017). https://doi.org/10.1021/acsami.7b03987
31. Humood, M., Qin, S., Song, Y., et al.: Influence of graphene reduction and polymer cross-linking on improving the interfacial properties of multilayer thin films. ACS Appl. Mater. Interfaces **9**, 1107–1118 (2017). https://doi.org/10.1021/acsami.6b13209
32. Deline, A.R., Frank, B.P., Smith, C.L., et al.: Influence of oxygen-containing functional groups on the environmental properties, transformations, and toxicity of carbon nanotubes. Chem. Rev. **120**, 11651–11697 (2020)
33. Wang, M., Yu, S., Wang, C., Kong, J.: Tracking the endocytic pathway of recombinant protein toxin delivered by multiwalled carbon nanotubes. ACS Nano **4**, 6483–6490 (2010). https://doi.org/10.1021/nn101445y
34. Deguchi, S., Alargova, R.G., Tsujii, K.: Stable dispersions of fullerenes, C60 and C70, in water. Preparation and characterization. Langmuir **17**, 6013–6017 (2001). https://doi.org/10.1021/la010651o
35. Lyon, D.Y., Fortner, J.D., Sayes, C.M., et al.: Bacterial cell association and antimicrobial activity of a C60 water suspension. Environ. Toxicol. Chem. **24**, 2757–2762 (2005). https://doi.org/10.1897/04-649R.1
36. Lyon, D.Y., Alvarez, P.J.J.: Fullerene water suspension (nC60) exerts antibacterial effects via ROS-independent protein oxidation. Environ. Sci. Technol. **42**, 8127–8132 (2008). https://doi.org/10.1021/es801869m
37. Kang, S., Mauter, M.S., Elimelech, M.: Physicochemical determinants of multiwalled carbon nanotube bacterial cytotoxicity. Environ. Sci. Technol. **42**, 7528–7534 (2008). https://doi.org/10.1021/es8010173
38. Vecitis, C.D., Zodrow, K.R., Kang, S., Elimelech, M.: Electronic-structure-dependent bacterial cytotoxicity of single-walled carbon nanotubes. ACS Nano **4**, 5471–5479 (2010). https://doi.org/10.1021/nn101558x
39. Huang, L., Xu, S., Wang, Z., et al.: Self-reporting and photothermally enhanced rapid bacterial killing on a laser-induced graphene mask. ACS Nano **14**, 12045–12053 (2020). https://doi.org/10.1021/acsnano.0c05330
40. Chong, Y., Ge, C., Fang, G., et al.: Light-enhanced antibacterial activity of graphene oxide, mainly via accelerated electron transfer. Environ. Sci. Technol. **51**, 10154–10161 (2017). https://doi.org/10.1021/acs.est.7b00663
41. Gurunathan, S., Han, J.W., Abdal Dayem, A., et al.: Oxidative stress-mediated antibacterial activity of graphene oxide and reduced graphene oxide in *Pseudomonas aeruginosa*. Int. J. Nanomed. **7**, 5901–5914 (2012). https://doi.org/10.2147/IJN.S37397
42. Hu, W., Peng, C., Luo, W., et al.: Graphene-based antibacterial paper. ACS Nano **4**, 4317–4323 (2010). https://doi.org/10.1021/nn101097v
43. Lee, M., Yang, S., Kim, K.J., et al.: Comparison of the catalytic oxidation reaction on graphene oxide and reduced graphene oxide. J. Phys. Chem. C **118**, 1142–1147 (2014). https://doi.org/10.1021/jp411682b

44. Musico, Y.L.F., Santos, C.M., Dalida, M.L.P., Rodrigues, D.F.: Surface modification of membrane filters using graphene and graphene oxide-based nanomaterials for bacterial inactivation and removal. In: ACS Sustainable Chemistry and Engineering, pp. 1559–1565. American Chemical Society (2014)
45. Xin, Q., Shah, H., Nawaz, A., et al.: Antibacterial carbon-based nanomaterials. Adv. Mater. **31** (2019). https://doi.org/10.1002/adma.201804838
46. Qie, X., Zan, M., Gui, P., et al.: Design, synthesis, and application of carbon dots with synergistic antibacterial activity. Front Bioeng. Biotechnol. **10** (2022). https://doi.org/10.3389/fbioe.2022.894100
47. Chu, X., Zhang, P., Wang, Y., et al.: Near-infrared carbon dot-based platform for bioimaging and photothermal/photodynamic/quaternary ammonium triple synergistic sterilization triggered by single NIR light source. Carbon N. Y. **176**, 126–138 (2021). https://doi.org/10.1016/j.carbon.2021.01.119
48. Chen, S., Quan, Y., Yu, Y.L., Wang, J.H.: Graphene quantum dot/silver nanoparticle hybrids with oxidase activities for antibacterial application. ACS Biomater. Sci. Eng. **3**, 313–321 (2017). https://doi.org/10.1021/acsbiomaterials.6b00644
49. García Vélez, D.G., Lagos Álvarez, K.J., Romero Obando, M.P.: Antibacterial Strategies: Photodynamic and Photothermal Treatments Based on Carbon-Based Materials. IntechOpen (2023)

# Chapter 6
# Conclusion, Challenges, and Future Perspectives

CBNMs are essential materials for tackling environmental concerns through various uses. Activated carbon and CNTs are very efficient adsorbents used in pollution remediation to remove contaminants from air, water, and soil. These materials may selectively absorb various contaminants, such as heavy metals, organic pollutants, and emerging contaminants, due to their huge surface areas and specific surface chemistries. This process helps to clean environments and protect human health. CBNMs are crucial in water treatment procedures, being used in filtration membranes and desalination technologies to ensure access to clean and safe drinking water. Gr-based membranes demonstrate outstanding water permeability and salt rejection characteristics, making them a potential option for effective water purification and desalination. CBNMs provide sustainable energy solutions by aiding in the advancement of renewable energy technology. These materials provide effective energy storage and conversion, lowering dependence on fossil fuels and decreasing greenhouse gas emissions. They are used in carbon electrodes for lithium–ion batteries and as catalysts in fuel cells. The environmental uses of CBNMs highlight their flexibility and effectiveness in addressing pollution, preserving natural resources, and promoting environmental sustainability for future generations.

## 6.1 Challenges

CBNMs have great potential for environmental applications but encounter several hurdles that need to be overcome for their general acceptance and efficacy.

- CBNMs, especially those designed at the nanoscale, can be costly to mass-produce. Expanding manufacturing methods while ensuring quality and consistency is a major obstacle, limiting their extensive use in environmental applications.

 
M. Ikram et al., *Carbon-Based Nanomaterials for Environmental Applications*, Engineering Materials, https://doi.org/10.1007/978-3-031-59390-1_6

- CBNMs need to retain their structural integrity and performance for lengthy durations to ensure their long-term usefulness in environmental applications. It is crucial to tackle problems with the degradation, clogging, and release of carbon-based materials in environmental settings to ensure their sustainable utilization and implementation.
- Some CBNMs, especially synthesized nanomaterials, may pose significant health and safety problems to humans and the environment. Evaluating and reducing any dangers linked to the synthesis, manipulation, and elimination of CBNMs are crucial to guarantee their safe and conscientious utilization in environmental contexts.
- The synthesis of CBNMs often require intricate and energy-demanding procedures, potentially leading to environmental harm and resource exhaustion. Creating sustainable and scalable methods of synthesis is essential for minimizing the environmental impact of producing CBNMs.
- Surface functionalization of CBNMs is necessary to improve their capacity to adsorb, select, and maintain stability for targeted environmental contaminants. Optimizing surface functionalization while preserving material integrity and performance is difficult, especially in changing and severe environmental conditions.
- Challenges arise about the cost, efficiency, and environmental effect of regenerating and reusing CBNMs following pollutant adsorption. Creating effective methods and plans for regenerating CBNMs is essential for optimizing their environmental and economic advantages.

## 6.2 Suggestions

To address the obstacles encountered by CBNMs in environmental applications, collaborative efforts are required on several fronts. Increased investment in research and development is essential to progress synthesis processes, surface functionalization technologies, and production scalability. Addition of metal, non-metal, and particularly polymers to CBNMs increased surface area, stability, and exciton separation, which is effective for photocatalytic dye degradation and antibacterial applications. Advanced characterization methods will offer more profound understanding of material characteristics, assisting in the development of carbon-based materials with improved performance and stability. Optimizing processes and improving efficiency can increase manufacturing cost-effectiveness and reduce environmental impact. Furthermore, creating precise regulatory frameworks and standards for safety and sustainability will guarantee responsible material production and utilization. Education and awareness campaigns will be crucial in promoting comprehension and backing for CBNMs among stakeholders. By utilizing a multidisciplinary approach and following these ideas, we may overcome hurdles and fully utilize CBNMs for environmental purposes, leading to a cleaner and more sustainable future.

The manufacturer's authorised representative in the EU is Springer Nature Customer Service Centre GmbH, Europaplatz 3, 69115 Heidelberg, Germany. If you have any concerns regarding our products, please contact ProductSafety@springernature.com

Printed and bound by CPI Group (UK) Ltd, Croydon, CR0 4YY
28/09/2026
02224074-0015